DISCOVERING OUR WORLD

DISCOVERING OUR WORLD

Humanity's Epic Journey from Myth to Knowledge

Paul Singh
and
John R. Shook

Illustrations by Matt DiPalma

PITCHSTONE PUBLISHING
Durham, North Carolina

Pitchstone Publishing
Durham, North Carolina 27705
www.pitchstonepublishing.com

To contact the publisher, please e-mail info@pitchstonepublishing.com

Printed in the United States of America

10 9 8 7 6 5 4 3 2 1

Library of Congress Cataloging-in-Publication Data

Singh, Paul, 1960-
  Discovering our world : humanity's epic journey from myth to knowledge / Paul
Singh and John R. Shook ; with illustrations by Matt DiPalma.
      pages cm
  Includes bibliographical references and index.
  ISBN 978-1-939578-14-3 (pbk. : alk. paper)
  1. Cosmology. 2. Philosophical anthropology. 3. Civilization. I. Shook, John R.
II. Title.
  BD511.S56 2014
  113—dc23
                        2014007703

CONTENTS

PREFACE

For the first time in humanity's existence, we possess solid information about the deepest questions that have puzzled our species. How did everything come to be? What is our place in this universe? Why does the Earth have its features? How did the Earth's forms of life arise? Why are human beings so similar, yet so different? What shall determine our destiny? Science now supplies firm knowledge about most of the crucial turning points in our deep history, the momentous events shaping who we are and what we may become.

This book explores many of the greatest questions humanity can ask, about the universe's long history and humanity's development to our own times. Its chapters contrast traditional tales of various religions with current scientific theories. Humanity's best knowledge is represented here, in brief explanations of ways that science came to understand so much about the universe's expansion from the Big Bang, the formation of the Earth and its special features, the origin and evolution of life, the origins of humanity, the development of human sociality and morality, the inventions of agriculture and civilizations, the political and cultural contests among empires, the collusion between monotheism and monarchies, the rise of Western civilization and its formative structures, and the development of science's knowledge and its applications in transforming technologies.

Science's exciting answers to the biggest questions about the world should be celebrated fervently and frequently. As a thoroughly human pursuit, science need not be confusing or alienating. By assembling a "big history" account of science's answers, we can be better enlightened about answers to questions such as "where did it all come from" and "how did life get here?" to "what makes us the way we are?" and "what may be the consequences of humanity's impact on our planet?" This book especially builds a great case for science itself, as readers are both intellectually and emotionally engaged by science's courageous journey and hard-won accomplishments.

Although science deserves all the credit for our abundance of knowledge, much of humanity is not ready to respect it. Preferring older myths enveloped in religious practices designed more to enslave than enlighten, billions of people resist science. They are unprepared to appreciate a radically new kind of story about the world. There is no place for gods and demons in science, nor can there be a privileged place for humanity within science's account of everything. All the same, the story science has to tell about the world is incomparably more surprising and interesting than any mythical tale about some clash of the titans or some theological system about a calculating creator.

Does science disenchant the world, leaving it cold and meaningless? Science does wake people from pleasant dreams, but the real world of science engages our entire wakened mind, making life only more meaningful. If science was just a collection of endless stale facts, pinned to textbooks like dead butterflies in a collector's case, people could excusably be repelled by such a lifeless display.

Science actually is a living, dynamic, and exciting thing. Besides science's story of the universe's evolution and our humble place in it, science has another story to tell, a narrative about itself. Let science tell its own story, a story of bold exploration, risky venture, brave confrontation, and glorious victory. Science should not be humble. Humanity may have no special place in the universe, but humanity is truly special because of our ability to comprehend our universe. How did the universe evolve to the point where a miniscule part of it could comprehend the rest?

This book is a celebration of science—of science's knowledge of the world, and of science's own journeys to gain that knowledge. More people should appreciate not only what science has to say, but also what science had to go through to be able to say it. Our cognitive processes have both rational and emotional aspects. It's in basic human psychology: we learn best from narratives, from stories that we feel involved with personally. Presented as both a grand narrative about the world, and as a magnificent narrative of human adventure, science can entrance, entice, and ennoble us. The accusation that science strips us of all significance is mistaken. Only the pleasant dreams of myth and legend now linger to undermine human intelligence. But humanity's better nature, eager for mental empowerment and involvement in something much bigger than one's self, can be enlisted in the journey of science, for that journey truly is humanity's journey.

1

THE ORIGIN OF THE UNIVERSE

In the beginning... How many stories have started that way? For as long as humans have gazed up toward the sky, they have wondered where it all came from. Their stories have become as uncountable as the stars. Wherever you go, anywhere in the world from desert to plain to valley or mountain, you will meet people who have lived there a long while and can tell you how it began. And not just where their own land or tribe came from. Talk to people a while and eventually you will be treated to some story about how everything began, all of it, everything from the Earth and the seas and even the stars, and especially about where all humans came from and why we are here.

We are storytelling animals. Perhaps we aren't the only species capable of language. But we are the only kind of animal that would make up a story and then believe it. Humans are a species that must have stories—we cannot learn how to live without our stories. There is something about the great size of our brains that gives us not only the capacity to talk with each other, but also the drive to figure out where we came from, and where we are going. As a species, we have been living by our legends for a very long time. Before there were scientists, before there were priests, before anything was written down, there were people who had plenty of time to watch the glorious skies. They wondered why there are stars shining down on the land, and why there are people gazing back up at the stars. And they began to tell stories.

Stories of Creation

Stories continued to be told as the first civilizations arose around six thousand years ago. From Egypt and Mesopotamia to India, China, and Central America, every gathering of peoples into the first small cities was accompanied by some organizing of the best stories. When writing was invented, many of the first things recorded on papyrus or clay or animal skins were the peoples' origin stories. We now regard those stories as those civilization's creation myths, containing imaginative ideas about the world that may be more than ten thousand years old or even older.

By the time that the world's first civilizations arose and flourished, people had long been sharing and trading their stories right along with everything else they had of value. Nearby cities borrowed and combined stories and neighboring civilizations soon had religions that shared as many similarities as differences. After the great trade routes across land and sea were able to connect the Mediterranean and the Middle East, and the Middle East with Central Asia and India, and India on with China, religious stories traveled with the travelers. The great civilizations, such as the Persian and Roman empires and the Gupta Empire in India, contained within their borders many religions simultaneously, and the blending and merging of mythological stories accelerated. While the names of gods and details of their deeds still proliferated, all those religions included similar themes and narrative plots.

The ancient Egyptians had several creation stories, dating back before 3000 BCE. Many involve the primeval and oldest kind of being, a fundamental watery darkness called the Nun. Vaster than the world, comprising what exists beyond the stars above and beneath the Earth below, Nun generated everything else alive, including the other gods. Egyptian myths add a rivalry between this watery deep and the first male god, usually associated with the sun. One myth important to the Heliopolis theology of the Old Kingdom (roughly 2600 to 2100 BCE) credits the supreme deity Atum with creating the world. In the beginning Atum was submerged among the chaotic waters called Nu or Nun. By first creating a hill of earth in the middle of the waters, Atum had a place to stand so he could then make two more gods, Shu and Tefnut, out of himself. Shu and Tefnut gave birth to the earth god Geb and the sky goddess Nut, who in turn birthed more gods including Osiris and Isis.

The earliest Babylonian creation story is preserved in the Enuma Elish clay tablets. In the beginning there were two gods. Apsu was the fresh water of the world, and Tiamat was the salt water (the oceans) of the world. Their intermingling produced more gods, such as a sky god and an earth god. After one of these gods, named Ea, kills Apsu before Apsu could destroy them, Ea's son Marduk has to do battle with a vengeful Tiamat. By killing Tiamat and dividing her body into the earth and the sky, Marduk created the world. Marduk then creates humans from the body of Tiamat's husband and places them on the earth to work as slaves to the gods.

Creation stories from the ancient Hebrews were recorded in Genesis, the first book of the Jewish Torah. As Genesis opens, in the beginning the gods called Elohim existed along with darkness and waters. (The early Hebrews were polytheists, and the singular god Yahweh's creation story starts in the second chapter of Genesis.) Deciding to create the Earth, the Elohim first created light and then they divided the waters to make room for sky, where they placed the sun and moon. By dividing the water under the sky, the earth was formed between the seas, and then the Elohim filled land and sea with living creatures. Finally the Elohim announced the creation of mankind, "in our image," and gave humans dominion over all the life on Earth. This creation tale is an example of how tribal communities residing near earlier great civilizations borrow from their older mythologies. The primary events of Genesis (dating from 800–500 BCE) echo much older myths commonly told around the Near East at least 3,000 years before Christ and 1,500 years before Moses supposedly wrote down the Torah. God had to create light and earth but not water, repeating a much older mythic idea found from Egypt and Mesopotamia to Persia and India associating water with a dangerous cosmic power which had to be controlled and divided.

Both ancient Egypt of the Nile and ancient Sumeria of Mesopotamia (now present-day Iraq) looked to vast cosmic waters as the stage where the drama of the world's creation took place. This made some sense for settled peoples relying on large rivers like the Nile and the Euphrates for irrigation farming. There's no reason for nomadic herders from the dry uplands to think much about huge rivers and vast floods, yet the Hebrew genesis story included not one, but two flood stories. In the beginning, God had to begin creation by dealing with the primeval and chaotic waters that were already there, dividing them just like Marduk had to divide Tiamat. The Hebrews were also familiar with a Sumerian story, also adopted by the Babylonians a millennia later, about a great flood striking the population centers, and so the story of Noah and the Flood entered the Hebrew genesis legends.

Ancient Hinduism is based on the Vedic scriptures. In the Rig Veda, a variety of hymns about the creation of the world are narrated. In the beginning there was only darkness and water. A Golden Embryo came into being by itself, which then generated the sky, the earth, and the sun. A later sacred Hindu text titled The Brahmanas credits the waters with creation. In the beginning the primeval waters formed a golden egg, from which came the first god named Prajapati. From Prajapati's breath came the light, the sky, and the earth, along with other gods.

The Yoruba have long inhabited a region of West Africa, today mostly in the country of Nigeria. According to one of the Yoruba creation myths, in the beginning only chaotic waters exist. The supreme god Olódùmarè (also called Olorun) sent a godly assistant named Obatala to make land. Obatala descended down to the waters on a chain, carrying a shell holding some earth, some iron, and a chicken. Obatala piled the iron in the water, then mounded the earth on top of the iron. When Obatala set the chicken down on the earth, the chicken scratched and scattered the earth around, creating all the land. After some minor gods had lived on the land awhile, it had sufficiently dried out, so Obatala then fashioned humans from some of the dry earth and Olódùmarè made them come to life by blowing into them.

Among the Finnish legends are tales preserved in the Kalevala, a collection of epic poetry. In the beginning there were only the waters and the sky. Sky had a daughter, Ilmatar, who let a beautiful bird make a nest on her knee. When the eggs broke, shells made land, the egg whites made the moon and stars, and a yolk became the sun. Ilmatar continued creation in the new land, forming the features of the earth. A son, born of her and the sea, was the first human.

A long poem by Hesiod called Theogeny records ancient Greek myths. In the beginning there was only Chaos, but two gods emerged in this chaos, Gaia of earth and Eros of love. Gaia and her children in turn produced further generations of gods, whose battles resulted in the eventual victory of Zeus and the final pantheon of Olympian gods. Along the way, humans get created only to end up as the playthings of the gods for their amusement.

Almost every culture includes legendary stories about how the world began from some original condition, how the first creator(s) did their mighty deeds, and how humans were formed along the way. Collections of creation myths can include hundreds of these sorts of stories, and even a one-volume survey such as *A Dictionary of Creation Myths* by David and Margaret Leeming offers a bewildering variety from all around the world.

Scholars of world mythology have been able to largely agree on some type of order for categorizing creation stories. Obvious similarities among legends can't be missed. For example, the presence of chaotic waters in the beginning keeps recurring across many cultural myths. Both unformed chaos, and the waters that symbolize its unpredictability, struck many curious minds as an obvious place to start from. The founding of the world is linked with the origin of all order and regularity and law, which must survive in competition with its chaotic origins. The motif of birth is also essential to many creation stories, and the story of how the first birth took place supplies "the beginning" to it all. From this first birth, either of the world or of a god who then makes the world, everything else is subsequently born. Birth suggests reproduction, and mythologies are filled with accounts of divine parents making more gods and sometimes humans as well.

Some creations myths are focused on ancestry. They start from humanlike beings with divine powers, and proceed on to recount their great deeds and the noble births of the next generations. Other creation myths mention ancestors, but they also try to explain how those ancestors came into existence. Perhaps they were born from the supreme god or gods, or perhaps the first people were made by the gods. And where did the gods come from? Some creation myths describe the origin of the first god out of something else, while other creation myths assume that the first god had always existed right along with the primordial chaotic conditions. Only a few creation stories, like those of Christianity or Islam, declare that only one true god existed from all time and then made everything else.

What seems to be quite common throughout so many creation myths is that creation cannot be explained by chaos, by randomness, or by ignorance. At some point in the story a creator has to get going and take charge. This creator is an agent, a being similar to a person with a mind and a plan, who is responsible for creating us humans and the world we inhabit. Different creation myths attribute quite different roles to the supreme god or gods. Sometimes the supreme gods are like parents, trying to get along well enough to be lovers and then parents, reproducing a divine family and birthing humanity along the way. Sometimes the gods are like benevolent lords, taking responsibility for establishing and managing the orderly world so that humans can have a hospitable place to live. Sometimes the gods are like cruel masters, creating humans only to immediately demand their obedience and loyal service. Sometimes the gods are like military generals, gathering humans into an army to fight in some divine war.

Whatever the general theme of a creation myth, you can be sure to find a divine agent or two involved. Things have come to be the way that they are because some god or gods have deliberately caused it to be that way. Creation myths try to explain how order came from disorder, how the first birth happened and what god was born in the process, how this god is responsible for the rest of creation, and the way that the first people were created in the process. Narratives in general are like that; the point of stories has always been to recount what people have been doing, and why they are doing what they do. The focus of mythological stories is always on the deeds of the creators, what they have accomplished, and how humanity plays some small role in the cosmic drama.

The stage for these dramas has never been as grand as the plots, though. The earth, the sky, the stars—that's usually as far as the religious imagination extended. The most ambitious of the ancient myths usually only manage to put the earth at the center of it all, closely surrounded by the vault of the sky and its embedded heavenly lights. The notion that the heavens are much farther away than the clouds, or that some stars are much farther away than the sun, never occurred to mythmakers. Only after early astronomy began realizing how far away the moon and the sun and the planets must be, did mythmakers start to update their visions. By medieval times, creation had become a little bigger, but the earth still occupied center stage, the heavens were still pretty close, and the whole of creation was depicted as a sphere with a definite center and a surrounding edge.

During the Renaissance, a few brave astronomers starting with Copernicus and Galileo dared to say that the earth was not at the center of everything. They instead suggested that all the planets go around the sun and that the stars might be distant suns with their own planets. But religions were not interested. Removing the earth and its people from the center of everything threatened to ruin a good story.

The Science of Starlight

The stars turned out to be better storytellers than us. Information from all those stars, near and far, permitted astronomers to figure out what the universe was actually like. The progress of astronomy delivered more amazing discoveries after it displaced the earth from the center of creation. Not only is the earth just one of several planets orbiting the sun, but the sun is just one of many billions of stars going round and round in a disc-shaped galaxy. During the eighteenth and nineteenth centuries, the best astronomy could do was to depict our galaxy of stars as the entire universe. This was big progress, at a time when most religious people still faithfully believed that the earth is the center of everything. Then the little globs of fuzzy light scattered between the stars turned out to be other galaxies far beyond our own galaxy, as discovered by Edwin Hubble using the largest telescopes available in the early 1920s. Billions of galaxies could be seen through bigger and bigger telescopes built later on during the twentieth century, and telescopes in orbit around the earth, such as the Hubble Telescope, have focused in on the most distant galaxies.

But the largest surprises were still coming. Not only did the universe contain many billions of galaxies and trillions of stars, far vaster than any religious picture, but the universe itself was completely different than anyone had imagined. Hubble next discovered that most of the visible galaxies around us were getting farther away from our own galaxy. By 1970 or so, astronomers had concluded that according to the best available evidence, the universe is incredibly huge, has no center and no edge, and had a small beginning about fourteen billion years ago. The universe was created from a tiny burst of intense energy, all the energy the universe would need from then on, in order to inflate to its current size and keep on inflating. Of all that energy, only a tiny fraction ended up as stars and planets, just thin debris left scattered around the universe. Our own place in the universe turned out to be completely insignificant.

The stars that we can see in the night sky made it look like we are at the center of everything. Telescopes allowed us to figure out what is really going on out there, especially after the galaxies were discovered. It turns out that there is no good evidence that we occupy any special place in the universe. However, the galaxies might have deceived us too, just as the stars already had deceived prescientific minds. From the perspective of our galaxy, the Milky Way, we can observe many millions of other galaxies. They are very far away, and they are all getting farther away from us. Why would all the other galaxies look like they are moving away from us, if we aren't at the center of the universe? It turns out that there must be a different explanation for what appears to be happening. We can tell how far away other galaxies are and how they are moving by carefully examining the light coming from all those galaxies.

It does look like all the universe's galaxies are getting farther away from each other. In the past, they were closer together. In fact, the universe is about 13.8 billion years old, so back then, all the energy contained in those millions of galaxies was packed closely together. Did the entire universe have its energetic birth here, shooting off in all directions from where we are now? Probably not. The special place where all light and matter came from should look very different. You'd expect that the place where such an immense amount of energy exploded outwards should be extraordinary. But as we look around the universe, stars and galaxies aren't much different from each other.

Our place in the cosmos certainly isn't anything special at all. Our star is like millions of others, drifting along as our galaxy slowly rotates. Our galaxy is a pretty typical galaxy, similar to millions of galaxies randomly strewn about the universe. If we were instead located at one of those other galaxies, we would observe the same distribution of galaxies around us. From that perspective, we could also see how the rest of the galaxies are moving away from that location too, at the same rate. Every occupant of any galaxy might suppose that they are located at the center of the universe, and they would all be wrong. The only logical option left is to conclude that there is no place in the universe where "it all began." There cannot be any center to the universe. All those galaxies aren't really moving away from one special central place. They do look like they are moving away from us, but some other explanation is needed. If those galaxies are not moving outwards, going farther and farther out into empty space, then they aren't really moving much at all. Still, the distance between galaxies is evidently increasing.

Astronomers can tell that other galaxies are getting farther away from us and from each other because light from distant galaxies is "redshifted" into longer frequencies. Light has frequencies, and when light is detected by the human eye, longer frequencies look redder, while shorter frequencies look bluer. We can observe how the light from distant galaxies looks redder, and the farther away a galaxy, the redder its light looks. What is happening to all that light, to make it look redder and redder, with longer and longer frequencies? We do know that light from distant galaxies takes a long time to get to us. We can observe a galaxy that is one billion light-years away only because light from that galaxy took about a billion light years to reach us.

Measuring very long intergalactic distances in terms of light years in an expanding universe requires some careful terminology. There is no clear definition of distance in which the distance to a galaxy that emitted its light, say one billion years ago, is one billion light years, since that simplistic equivalence of distance with time ignores the expansion of the universe. There are several different distances used in modern cosmology, for example, metric distance, luminosity distance, and angular diameter distance, each used in different contexts. In this chapter, we give metric distance as the distance to a distant galaxy where the object is "now"—that is to say when a clock starts at the Big Bang and moves with the galaxy reads the same time as a similar clock here in our own galaxy, about 13.8 billion years. The matter that emitted the cosmic background radiation about 350,000 years after the Big Bang is now about 47 billion light years away from us, and moving away from us much faster than the speed of light. This does not contradict relativity—indeed this calculation is according to General Relativity—because it is space-time itself that is expanding, and nothing within space-time is moving so impossibly fast. The universe has expanded a little more than a factor of 1,000 since the cosmic background radiation was emitted; hence, this matter was only about 47 million light years away from the matter that would become us when that light was emitted.

Now, light is light, regardless of how old it is; light doesn't "age" or "get tired" or decay over time. Time can't change light's energy, but distance can make that energy look different. Light is made of photons that have frequencies, like pulsing waves of energy. Forcing a photon to travel through an increasing distance has the effect of "stretching" its energy, and that stretching causes the photon to have a longer frequency. Imagine two people holding a rope and waving it up and down fast enough to make the rope wave up and down. If one of the people takes a step backwards, the rope's waves will get longer, and now the only way to keep the rope's waves the same is for that person to put more vigorous energy into their end of the rope. But if no extra energy is put into the rope, its waves always get longer if the two people take steps farther apart. Unlike that rope, there is no way for a photon's waves to acquire any extra energy as it travels through space.

If a photon is traveling through an increasing amount of space, then its frequency must get longer. Therefore, light would have longer "red shifted" frequencies only if it had to pass through an increasing distance to make it all the way to us. The same effect of increasing distance on sound is called the Doppler Effect.

The Expansion of the Universe

We also can notice how the farther away a galaxy is from us, the more its frequency is red shifted. For example, light from galaxies twice as far away has about twice the amount of red shifting. The galaxies the farthest away from us are getting farther away at the fastest rate. That means that those galaxies aren't all speeding away from us as if they started close to us in the past and they have been moving away from us ever since. If all the other galaxies had started out near us and have been moving away from us ever since, we would expect to see them all still moving away from us at about the same rate.

By analogy, if we stood at the starting line of a race but let the other sprinters run off down the track, they would be moving away from us at about the same pace. The pack of average runners would be able to race away from us at about the same speed. After a minute, there would be a lead runner in the pack, a slowest runner at the back of the pack, and no runners at all between the last runner and us back at the start. But when we look at the galaxies around us, they aren't moving away in a pack. They are all spreading out away from each other. They aren't like runners in race. They are more like raisins embedded in an expanding loaf of bread. Starting out close together in a lump of dough, the raisins would get farther away from each other as the dough is slowly baked into a much bigger loaf. Raisins that began right next to each other would soon have growing gaps between them. Two raisins that started out an inch from each other would be two inches and then four inches apart. From the perspective of each raisin, the other raisins would look like they are getting farther away, and the farthest raisins would seem to be moving away the fastest. This analogy is described in Martin Gardner's 1962 book *Relativity for the Million*.

Our universe is more like an expanding loaf of bread, with each galaxy behaving like a raisin embedded inside that loaf. In that loaf, no raisin is actually moving around inside the growing loaf of bread at all. The universe's galaxies aren't speeding around inside it, either. Galaxies clump together a bit because of their gravitational pull on each other, but the distances between clusters of galaxies is still increasing nonetheless, and gravity is not enough to hold all the universe's galaxies together. None of the galaxies have to move in order to get farther away from each other. The universe itself has been getting larger ever since its origin from a very small beginning, the "Big Bang."

There is an important way that the universe is not like a loaf of bread. If a raisin at the surface of the loaf looks around using light, it would see all the other raisins on one side of it, and no raisins on the other side, out beyond the loaf's crust. However, we can't see any "crust" or edge to the universe, and none of the other galaxies would spot an edge either. From the perspective of every galaxy, it looks like it is sitting in the middle of the loaf, not near any edge where the galaxies stop. How could every galaxy seem like it is in the middle of the universe, if the universe is expanding outwards? Wouldn't there be an outer surface to the universe, and couldn't the galaxies near that surface be able to peer farther out to see the empty space beyond the universe? The problem is that there is no galaxy in any position to get a glimpse of anything beyond the universe, because all light arriving at this galaxy exists within the universe's own space. The galaxy could only see what is already inside the expanding universe. By analogy, suppose those raisins can detect each other only through vibrations traveling through the starchy bread between them. Could a raisin detect anything out beyond the bread? Of course not. Everything that this raisin could observe would come from inside that loaf, and as far as that raisin could tell, there is no edge to its starchy universe. It couldn't ever become aware that it was on the crust of the loaf.

Like that raisin, we cannot detect any edge or surface to the universe because all the light arriving at our position travels within the universe itself; there is no light arriving from beyond our universe and no absence of light either. We cannot see any place in the universe where the universe stops and some other kind of space starts. We only see more and more galaxies everywhere we look. The very space of the whole universe is expanding: the universe is made up of both the three-dimensional space and all the matter and energy within it. We have grasped why it can't be the case that the universe's galaxies are expanding outwards into empty space that already existed before the Big Bang. That can't be, because the universe's space originated along with everything else in it, in the initial explosion of the Big Bang. Space itself can be stretched out, which explains how the universe is expanding and very distant galaxies keep getting farther away.

Furthermore, even when we glimpse very distant galaxies, we are seeing back into the past where the universe has been, not looking ahead where the future is going. When we observe a galaxy one billion light-years away, we see it as it was one billion years ago, since that was when its light left it to start its long journey to us. When we observe a galaxy 10 billion light-years away, we see it as it was 10 billion years ago. The oldest observable galaxy is just 13.1 billion light-years away, just a few hundred thousand years after the universe's origin, which nicely confirms other calculations of the universe's age. The farthest galaxies are also the earliest galaxies, forming quickly after stars themselves began to shine because of gravity's ability to concentrate gas clouds after the universe had a chance to expand. When we see what is far away, we see what was going on long ago.

What is the rest of the universe like now? Even though we can't see that entire universe as it is right now, the universe is the same age everywhere, so it is reasonable to suppose that all the other galaxies in the universe are doing pretty much what our galaxy is doing now. Our Milky Way galaxy is recycling its dying stars into new stars, absorbing nearby little clusters of stars, and slowly drifting into a squishy collision with the nearby large galaxy, Andromeda, about 4.5 billion years from now. From the perspective of every galaxy, the universe looks about 13.8 billion years old, and it looks pretty much the same in all directions. The universe is everywhere expanding, but it has no center and it has no edge.

Another way to tell that the universe has no edge is to notice how space itself is "flat" so that light everywhere goes in straight lines. As far as we can tell, light only bends when it passes close to some massive object like a star or a galaxy. When light is just passing through vast stretches of empty space, it goes in a straight line no matter what direction it is going. In the early history of the universe, light would not have gone so straight. When the universe was very small, soon after the Big Bang, space itself would have been "curved" and all light would have curved right along with it. But the rapid inflation of the universe eliminated that curvature, and that is why it is called "flat." We need a different analogy to help us picture the expansion of the universe. Imagine drawing a lot of little dots all over a rubber balloon before it is inflated. The dots will be very close together, because there is very little air in the balloon. As air is pumped into the balloon, the balloon expands, and all the dots get farther away from each other as the rubber of the balloon expands.

Ignore the empty air inside the balloon—in this analogy, the balloon is made up of just its rubber, just as the universe is made of its space, along with the dots embedded in the rubber, like the galaxies embedded within the expanding space of the universe. The dots "live" in a two-dimensional world, the two dimensions of the balloon rubber, and that is all those dots could ever detect. The balloon rubber is expanding in a surrounding three-dimensional space, but the rubber itself is not growing into any third dimension—it remains two-dimensional and so do the dots. There is no "center" to the balloon rubber, no place in the rubber where it began to expand. By analogy, our universe can be pictured as the three-dimensional surface of a growing "balloon" that is expanding in a wider fourth-dimensional space. As the universe "balloons" to an immense size after the Big Bang, it is not becoming four-dimensional itself, and no fourth dimension could be detected. There is no edge and no center to the universe, and no place in the universe where it began inflating.

The flatness of the universe is a strange concept, because we are used to only applying the idea of flatness to two-dimensional things. But flatness is a concept that can be applied to a three-dimensional thing as well. Going back to our balloon, when the balloon was still small, each dot could notice that the space around was curved because nearby light would not go straight. For example, any three dots make a triangle, and the light lines between them make the sides of the triangle. Do the three angles of that triangle add up to exactly 180 degrees, or do they add up to more than 180 degrees? If they add up to more than 180 degrees, then the triangle's space is curved, and indeed any triangle drawn on a small balloon will have angles that add up to more than 180 degrees. But suppose that this balloon has expanded to the size of the whole planet. Three dots within sight of each other will make a small triangle still, but that little area is so tiny compared to the entire surface of the balloon that its angles will add up to 180 degrees. The same thing is true for small triangles drawn on the surface of the earth. Even though the earth is a sphere, any triangle you could draw across your neighborhood would look perfectly flat with 180 degree angles. That is why we can use ordinary Euclidean geometry for drawing local maps, but airplanes have to take into account the earth's curvature when they travel long distances.

Returning to our universe, it is so immense that any "local" triangle will make 180 degrees every time. This has been observed to be the case by an examination of tiny variations in the cosmic microwave background, the oldest energy still observable, left over from soon after the Big Bang itself. Cosmologists refer to the universe as flat because light traces straight lines as it travels, excepting only when it passes close to some object with hefty mass, like a star or a whole galaxy.

The Birth of Our Universe

Our best evidence, most of our only evidence, about the way the universe began consists of light that did not come from stars. The original light of the universe, the cosmic microwave background still radiating everywhere, tells us much about the universe's origins as well as its shape and size. This "first light" is not made of the universe's first energies, because those energies were not in the form of photons, the units of radiation, and the earliest photons could not escape from the grip of all the other earliest particles of energy in the Big Bang. As soon as the universe expanded and cooled enough, when the universe was around 350,000 years old, to finally permit photons to set off on their speed-of-light journeys in all directions, all those photons began zipping across the universe—and have been doing so ever since.

We can listen to these echoes from the early universe on the radio, which interprets these first photons as signal noise, part of that static you can hear on an unused channel. About 10% of that static is from cosmic radiation, and the rest is from closer sources such as our sun, the earth below us, and (lately) other electrical devices around us. This radiation has very little energy in the extreme microwave range. During the 1990s and early 2000s specially designed satellites such as Cosmic Background Explorer (COBE) and Wilkinson Microwave Anisotropy Probe (WMAP) have recorded and measured this cosmic microwave radiation. One decisive confirmation that the universe's shape is flat was found by these instruments, when the small variations across regions of the cosmic radiation were shown to have just the right size consistent with universal flatness. And all this evidence from the universe's "first light" supplies our essential clues to the universe's origin.

Three unusual features of this cosmic background radiation have forced cosmologists to fit their theories about the universe's origins to that strange evidence. First, this radiation arrives at the earth from all directions in space in the same way. Second, this radiation is very faint but almost perfectly evenly distributed, so that the same amount of radiation arrives at the earth from every direction with only tiny variations. Third, all that radiation from every direction has almost the same energy. When cosmologists imagine what the universe was like at the 350,000-year stage of the universe, these three features raise a serious puzzle called the "Horizon Problem." How could all that "first light" from every part of that early universe have almost the same energy? The only explanation is that every part of that early universe had almost exactly the same density and energy as every other part. Such homogeneity raises an even worse puzzle: how could that early universe have such a uniform consistency? This puzzle is deeply connected with another puzzle about why the universe that we observe now is flat. It is so flat that light goes in straight lines (except when passing by large objects) everywhere and every part of the whole visible universe looks about the same. The trouble behind these puzzles is that there appeared to be no good reason why a young universe would have such a smooth and even consistency.

Because Albert Einstein tried to make sure that his relativistic equations for the universe's growth would guarantee a perfectly flat universe by arbitrarily adding a "cosmological constant," this puzzle is often labeled as the Flatness problem or the Cosmological Constant problem. During the middle of the twentieth century, two opposing camps of scientists disagreed about the reason why the universe is expanding the way that it is. The "steady state" group preferred the explanation that the universe has always existed but new space and matter is continually created throughout the universe, pushing everything apart in the process. The leader of this group, respected astronomer Fred Hoyle, ridiculed the other camp's view that the universe came from a tiny burst of huge energy by calling it the "Big Bang" hypothesis.

Hoyle's label of the "Big Bang" was soon adopted by everyone, but his rival theory was rejected. Far better evidence for the Big Bang soon accumulated, although fresh puzzles arose. A serious worry for the Big Bang theory arrived in the late 1990s, when it was discovered that the universe has not only been expanding, but this expansion is accelerating. The accelerating expansion of the universe seemingly required some undiscovered force, so cosmologists had to postulate something they called "dark energy."

Although the Big Bang is a plausible theory, it only explains some broad features of the observed universe and, by itself, leaves other features unexplained. Why was the early universe so evenly smooth and perfectly flat, with the right amount of mass and this "dark energy" which so perfectly balances the universe to ensure its growth? The early expanding universe didn't have to be so smooth. It could have been pretty lumpy and uneven, without anything to make it the same all the way through. The universe is not like a cake, made from batter that was mixed carefully and stirred thoroughly, and then allowed to cook evenly. The universe was rapidly expanding at the 300,000-year stage, from a much tinier size. Without enough time for any mixing or intermingling of energies to evenly distribute everything, the early universe shouldn't have been so homogeneous and flat. Something else must have been happening as the universe grew.

Alan Guth proposed a new theory about the very early universe in 1980, which he called the "inflationary model." According to this inflationary model, the early universe expanded extremely rapidly from an initial state of homogeneity. This period of inflation was very brief: just one billionth of a billionth of a trillionth of a second. But during this period, the universe expanded from subatomic size to perhaps the size of a marble or maybe as big as the earth. After this fast inflation period ended, at a moment that is properly labeled as the Big Bang, the universe settled into a slower expansion rate as an extremely even and smooth ball of high energy.

Because the universe right after the Big Bang was pretty much the same everywhere, with only minor variations sprinkled through the ball due to tiny quantum fluctuations, the Horizon Problem can be resolved. As the universe expanded and cooled after the Big Bang, it retained its mostly homogenous nature throughout. The light eventually sprayed by all parts of the early universe would therefore be expected to be almost exactly the same distribution and energy. This theoretical expectation has been neatly matched by the observed evidence from the cosmic background radiation. Those minor quantum fluctuations in the very early universe got much larger, expanding right along with the whole universe, providing the growing universe with immense ripples that later clumped galaxies together, just as we can now observe too.

An additional thing became clear about the universe after the inflationary model became well-recognized. The universe that we can observe, all that light from all those galaxies surrounding us, must be only a tiny fraction of the entire universe that came from the energy of the Big Bang. Cosmologists have to be careful to distinguish two universes to talk about: the entire universe and the observable universe. Most of the entire universe will never be observed by us. We can only observe distant galaxies if their light has been able to reach us, which started out in the early universe from some place close enough to us. We can see galaxies that are 13 billion light years away because there has been at least 13 billion years for light to get from that galaxy to the earth. The early universe made far more galaxies than just the ones we can now see, but they are in portions of the universe that expanded away from our portion. They must remain forever beyond the detection of any telescope, since their light can't ever get to us.

Because the universe's expansion is not slowing down, and actually appears to be accelerating, the visible universe can't get bigger. Indeed, it will get smaller and smaller billions of years into the future, because only light from closer galaxies will reach our galaxy then. Ironically, the visible universe will get smaller far into the future while the whole universe continues to expand. It is very difficult to estimate the total amount of the whole universe, which could be millions or billions times larger than our visible universe. The notion that there could be other universes besides ours has been partially confirmed, in the sense that there are many, many regions of the entire universe. Some parts could be quite similar to ours, with the identical physical laws and the same kinds of stars and galaxies. Other parts could be a little different, with minor or major differences in their physical laws.

Cosmologists cannot yet be sure what the whole universe must be like, because a theoretical explanation of the Big Bang has not been firmly established. One thing is sure, that the whole universe is so much more immense, that even if it still had just a little bit of curvature as a whole, our observable universe is such a tiny part that its own local curvature would look extremely flat. By analogy, maps of your town can depict its land as perfectly flat, because that land is a tiny region on the surface of a very large sphere.

The cause of the Big Bang still needs explaining, and so does the remaining Cosmological Constant problem. Where did all the initial energy of the universe come from, and why did that energy come in just the right amount to make a flat expanding universe? Although the inflationary model has explained why the early universe was so homogenous and flat, that brief period of inflation requires an explanation too. Over the past thirty years, many cosmologists have collaborated to produce the current model of the Big Bang itself. This basic model promises to explain the remaining puzzles, but it consists of many tentative and complex theories about particle physics, quantum mechanics, and new speculations such as string theory. Without going into technical details, this model relies on basic laws of physics and quantum mechanics. To explain the Big Bang, it is obviously necessary to postulate that something existed to be causally responsible for the Big Bang. Nothing could be more confusing than to suppose that science would try to explain how the Big Bang could happen from nothing at all. Cosmologists are busily proposing how the Big Bang could have been caused by an eternal realm of quantum chaos, black holes from a previous universe, or by collisions of other universes, and more theories will be invented in coming decades.

Cosmologists don't fully understand how the four fundamental forces of nature emerged from the cooling Big Bang. However, there is a degree of confidence in the idea that the Big Bang happened as those four forces were closely unified in one supreme force of original energy built up during inflation. As the universe continued to expand after the Big Bang, its internal energy became less dense and quickly cooled off, although it would have still been at 3,000 degrees after 300,000 years, permitting everything to become less fluid and more "lumpy." The basic forces and the particles which manifest those forces split apart and began interacting with each other, so that electrons, neutrons, and protons could form and then in turn attract each other to make the simplest atoms of hydrogen and helium.

When the universe was perhaps one or two hundred million years old, internal gravity lumped those gases into the hot clouds that became the first stars. Once the universe had made stars and their attending planets from the leftover dust around stars, only more time was needed. Later generation stars, made from earlier stars' explosions in supernovas, would shine down on evolved life forms that could gaze back at them.

Gods and Universes
Religions have imagined how gods made our world. So many gods, just to explain one world! Surely so many gods are not needed; religions can't all be correct. In fact, no religions are correct. Gods are not needed to explain our universe. However, we may be observing our universe to be the way it is because there are many other universes.

Cosmologists had hoped that our universe's actual laws of nature would be precisely specified by the best theory explaining the Big Bang. Although the inflationary model predicts that there would be vast energies obeying fundamental laws for the forces of nature, it does not predict exactly what those forces would be. Interestingly, the more plausible the inflationary model becomes, the less it is needed to make such exact predictions. Recall the quantum fluctuations that create false vacuums and the bubbles within them inflating while they try to return to the true vacuum. There is nothing in this inflation model or in quantum mechanics that says that there would only be a few bubbles. Indeed, there would probably be vast numbers of them, perhaps an infinite number. On this picture of inflation, sometimes called "eternal chaotic inflation," all these bubbles would be somewhat different from each other, so that every possible kind would get produced. According to the inflation model, a bubble that managed to produce a Big Bang and a resulting universe would have some narrow range of features, but among those growing universes, a sizable variety would have somewhat different sets of fundamental laws. Our universe may happen to be just one among innumerable kinds of universes, and there is no particular reason why it has the specific features that it does.

Our universe is not exactly random, nor did it come into existence from sheer chance. Just as it is a mistake to suppose that our universe came from absolute nothingness, it is a mistake to suppose that our universe's structure came from pure randomness. Many, but not all, of our universe's features had to be more or less the way that they are. First, as we have seen, our large universe came from a Big Bang, which means that it originated in some surviving portion of bubble that successfully inflated beyond the tiny quantum scale. Second, our universe's overall structure was dictated by that successful inflation and Big Bang explosion, and it probably has its gravity, its flat three-dimensional space-time, and its other basic forces because of that successful growth. Third, our universe expanded and lived long enough to produce hot stars and rocky planets capable of supporting life like us. We can now observe the universe the way it is because the universe matured to the point of being capable of producing us. In fact, we exist in the universe, at this 13.8 billion year stage, because that is within the universe's most fertile period for easily supporting life. Many generations of stars have come and gone in supernova. The death-stages of stars are times when the larger atoms like carbon and oxygen and nitrogen, necessary for organic molecules, get made and sprayed across the galactic neighborhood. During this age of stars, before they run out of fuel and the universe expands too fast, life has time to emerge, evolve, and look around.

It cannot be surprising that we find ourselves living in a universe that has grown and evolved to the point that it can support us. If the universe were just a little bit different, if the strength of gravity or another fundamental force were just a little weaker or stronger, for example, then it could not have supported life like us. Does this mean that the universe must be the way it is, because we exist? This is a tempting idea to some people, but it is not a scientific idea. Supposing that we are the most important thing in the world, and that the rest of creation depends on us, is a religious idea. The universe does not exist so that we can exist. If we were destined to exist, it is because our universe got around to making us, and therefore we are dependent on the universe, not the other way around.

The universe that astronomy can see is hardly designed for life, since far less than one trillionth of one trillionth of all the universe's vast space is even close to being warm and wet enough for life. Almost all of the universe, at least 99.999999999999999 percent of it, is as hostile to life as could be imagined. The immensity of intergalactic regions consists of almost absolutely empty and cold space only traversed by deadly high-energy radiation. Most of the rest of the friendlier warmer space within galaxies is populated by widely separated stars, a majority of which are either too cool or too hot, too short-lived, or so bound to multistar systems, forbidding any stable and habitable planets to orbit them. Furthermore, even the nicer pockets of life-friendly universe near comfortable stars will only last for less than one hundred billion years into the future, as the ever-increasing expansion of the universe tears apart galaxies and then solar systems and planets. One hundred billion years is a long time, but this sad fate of the universe is another reminder that the universe as a whole is not friendly toward life.

Even if ours were the only universe, science tells us that there is no good reason to decide that it was designed for us. As cosmologists grow more comfortable with the theory that ours is one of many, many universes that can have a wide variety of structures and laws, there is even less of a reason to think that our universe is special. Our universe is special to us, of course, since we inhabit it, but the universe did not have to be the way it is.

By analogy, we no longer assume that the Earth was made just for us. At one time, before we realized that there must be millions or billions of other planets just in our galaxy alone, it could seem to us that we lived on a truly special planet, so well-designed just for us. But now, after science has discovered the facts about the natural evolution of life on this planet and the vast variety of planets scattered around the galaxy, the Earth's existence no longer seems so mysterious or special. With all those planets, some just like the Earth would have been formed along the way, along with many more that permit different kinds of life. We just happen to have evolved to survive on this planet, and there's nothing special about this planet other than it happens to be able to support life like ours.

Just a diversity of universes, and the evolution of our universe, is enough to account for our existence. The natural processes that generate innumerable potential universes supply the vast abundance of variety so that a surviving universe just like ours gets inevitably created too. Once a universe like ours is created, its evolution towards creating innumerable stars and life-supporting planets provides the opportunity for life to get started on some of those planets. Other universes fairly similar to ours will likely contain intelligent life forms too, and they will also wonder if they are in a special universe. If they invent science too, they won't have to wonder for long. They can also leave behind many mythologies about gods creating one world according to intentional plans, and replace all that with one scientific account of a large collection of universes created according to physical laws.

The mythological mind can imagine all sorts of gods who planned and made one special creation for us. Such imagination is driven by the compelling wish that we are incredibly special, tremendously important, and occupy the most privileged place in the world. Mythological wish fulfillment is a powerful psychological thing, and intelligent people can fall victim to its seductive assurances.

When intelligence is twisted into the task of inventing comforting stories, the human imagination can imagine an innumerable number of "explanations" for the world. That is why there is, and always will be, the widest possible variety of religious narratives about so many gods assigned to the task of making one world. When the imagination is disconnected from having to rationally explain all of the actual evidence, it spins out innumerable stories all trying to explain one thing. This is good for religion in general, because people needing a wish-fulfilling confirmation of their own importance can always find some mythological tale that sounds just right to them.

Science is in the business of rationally explaining all of the evidence surrounding us, rather than egotistically confirming some special status

for us. Current cosmology has discovered that we occupy no special place in a universe that happened to be just unusual enough to support life like us. But being a little out of the ordinary is really nothing so special—not special enough to tie down intelligence with mythological fantasies misleading the mind with euphoric wish fulfillment. Science aims at intelligently understanding reality, whatever reality is there regardless of whether we wish it to be so.

Religion has always responded to science by claiming that there must be more reality beyond the nature than science can investigate. Obviously, cosmology does not yet have all the answers about how our universe was created. Among the immense questions still confronting science are, what existed before the Big Bang, and did time exist before our universe? Supporters of the Big Bang standard model sketched above suggest that the idea of time before the Big Bang is a meaningless concept, and asking about it generates meaningless questions. They are right in the sense that our idea of time is our own and makes sense only within our universe of space-time that was created when the universe began. Linear and continuous time may only exist on scales larger than subatomic scales but smaller than entire universes. There may be no linear time at the quantum level, and other universes would have their own separate stretches of time. Using our ordinary sense of time to think about what is happening with other universes makes as much sense as asking if there is Daylight Savings Time on Neptune.

Even after setting aside meaningless questions and their unnecessary answers, religions can still make people feel like their gods are easier to understand than the explanations from science. Indeed, the special events described during the first three minutes of the Big Bang are inherently unsatisfying to many people. Cosmologists are not confident that they fully understand those intense and complex natural processes involved in our universe's birth. With so many questions remaining unanswered, the current understanding of the Big Bang is surely flawed in ways that will be revealed by further research and experimentation. New experiments at research centers, such as the Large Hadron Collider at CERN in Europe, will hopefully reveal how the basic laws of nature relate to each and consequently clarify our universe's origin.

Religious people may complain that science is expecting us to irrationally think that the universe must have mysteriously come from nothing, and they offer a creator god to relieve such irrationality. But "nothing" isn't what is used to be. As already mentioned in earlier sections, followers of the standard model such as Stephen Hawking contend that it is possible for the universe to come out of "nothing," because in physics "nothing" is "not nothing," since it is something. The matter and antimatter which comes in and out of existence cancel each other out. It is a state of pure energy, so to speak. Physics has long been comfortable with the fact that matter (mass) can arise from pure energy and vice versa. The well-verified equation of $E=MC^2$ permits the universe's matter to come out of a potential energy that already existed. The prior conditions responsible for the Big Bang could be thought of as the universe's "creator," but appealing to such conditions is hardly an admission that there must be a creator god. A variety of cosmological theories speculate about how our universe could have been sparked into existence, and none of those theories describe anything that could possibly look like a religion's god.

Perhaps the idea of our universe getting "created" is one that arouses religious connotations for some, but the religions' various notions of gods can no longer get any comfort from science. Nature will likely prove sufficient to explain nature; it surely has so far, as Albert Einstein did, along with many other scientists and thinkers who came before and after him. Despite the way that science always has more questions to answer, such gaps in our knowledge should not allow us to fill those gaps with religious superstition instead of more science. We should continue to try to understand our world in reasonable ways and keep mythical beliefs out of the equation. The empirical methods of science have been our best guide for a long time. Religious ideas about our universe, by comparison, have been proven wrong innumerable times throughout recorded history. It should be embarrassing for anyone in today's modern world to let any religion dictate some origin story about our world.

Promoters of religion are also heard objecting how the scientific worldview can make us feel lost and abandoned without a god. But once again religion is trying to offer something that no one really needs. Letting go of an imaginary god living in some other world is the right opportunity for getting to know our real parent, the natural world. Why should we feel that we can no longer establish a relationship with our amazing universe? Learning more and more about our universe with the passage of time only makes us realize how deeply connected we are with the awesome and sometimes mysterious powers of our own world.

This is not a time to be disheartened. This is time to rejoice in our participation in the journey of learning about the breathtaking universe in which we live. Learning about our universe can only make us fall in love all over again, a love that will last for our entire lives and the lives of those who follow us. Our scientific discoveries have clearly shown us that our world is more amazing, awe-inspiring, and enchanting than anything religious prophets have ever been able to imagine.

Science can make us into deeply religious people, in a new natural sense of "religious" that refers to the inspiring insights of learning, not the blindness of superstition. The power of the scientific method should not be underestimated or misunderstood. Science has always been a modest but potent form of spiritual experience, engaging the greatest and fullest powers of the whole human mind, for those who experience and understand it. Understanding our true relationship with the universe is the only thing that can genuinely make us loving, kind, forgiving, and compassionate in our world that is so full of suffering. Religions have generated some very kindhearted people, to be sure, but so very few. Religious saints stand out in the shining light precisely because they stand against a religious backdrop of nightmarish darkness. Religion has created so much more conflict among us. Science has the potential to help us understand and love our neighbors and our beautiful world.

Further Reading

Bartusiak, Marcia. *Archives of the Universe: 100 Discoveries That Transformed Our Understanding of the Cosmos*. New York: Random House, 2010.

Bieri, Lydia, and Harry Nussbaumer. *Discovering the Expanding Universe*. Cambridge, UK: Cambridge University Press, 2009.

Christian, David. *Maps of Time: An Introduction to Big History*. Berkeley: University of California Press, 2011.

Couper, Heather, and Nigel Henbest. *The Story of Astronomy: How the Universe Revealed its Secrets*. London: Hachette, 2011.

Fox, Karen C. *The Big Bang Theory: What It Is, Where It Came From, and Why It Works*. Malden, Mass.: Wiley-Blackwell, 2002.

Gardner, Martin. *Relativity for the Million*. New York: Harper and Brothers, 1962.

Geary, Patrick J. *Women at the Beginning: Origin Myths from the Amazons to the Virgin Mary*. Princeton, N.J.: Princeton University Press, 2009.

Gleiser, Marcelo. *The Dancing Universe: From Creation Myths to the Big Bang*. Lebanon, N.H.: University Press of New England, 2005.

Greene, Brian. *The Hidden Reality: Parallel Universes and the Deep Laws of the Cosmos*. New York: Random House, 2011.

Kragh, Helge S. *Conceptions of Cosmos, from Myths to the Accelerating Universe: A History of Cosmology*. Oxford: Oxford University Press, 2007.

Krauss, Lawrence M. *A Universe from Nothing: Why There Is Something rather than Nothing*. New York: Simon and Schuster, 2012.

Leeming, David, and Margaret Leeming. *A Dictionary of Creation Myths*. New York: Oxford University Press, 1995.

Mather, John. *The Very First Light: The True Inside Story of the Scientific Journey Back to the Dawn of the Universe*. New York: Basic Books, 2008.

Montelle, Clemency. *Chasing Shadows: Mathematics, Astronomy, and the Early History of Eclipse Reckoning*. Baltimore, Md.: Johns Hopkins University Press, 2010.

Ruether, Rosemary Radford. *Goddesses and the Divine Feminine: A Western Religious History*. Berkeley: University of California Press, 2005.

Shaver, Peter. *Cosmic Heritage: Evolution from the Big Bang to Conscious Life*. Berlin and New York: Springer, 2011.

Van Voorst, Robert E., ed. *Anthology of World Scriptures*. Richmont, Cal.: Thomson Wadsworth, 2007.

Yankah, Kwesi, and Philip M. Peek. *African Folklore: An Encyclopedia*. New York: Routledge, 2004.

2

THE AGES OF PLANET EARTH

Delving into the big history of our beloved home that we call Earth provides us with a map that shows our own niche in space and time. A grand synthesis of disciplines of cosmology, geology, anthropology, biology, and big history has given us insights that we never imagined were possible just a few decades ago. In this chapter, we continue on our astonishing journey from Big Bang to the formation of the solar system, to the subsequent evolutionary history of our planet.

It is dizzying even to begin to think about deep time. One hundred years ago we were inventing the very first automobiles. Multiply that by ten, just a thousand years ago, imagine how most people believed that our Sun and planets were orbiting around the Earth and that we were at the center of everything that there is. The next factor of ten takes us back to the rise of earliest civilizations, inventing agriculture and trying to settle down from living a nomadic life. Another factor of ten, 100,000 years ago, witnessed the expansion of our own species across Africa not long after its evolution from hominin cousins now all extinct. Multiply by another factor of ten, when our hairy ancestors called Homo erectus walked around the Earth a million years ago. Just about 10 million years ago, our ancestors were living in trees as primates, during the great age of the mammals. One hundred million years ago, dinosaurs ruled the Earth, while small mammals were staying out of sight. A billion years ago, only single celled organisms thrived in Earth's oceans. Another multiple of 10 takes us back to a time when our solar system was yet to be born in our galaxy, the Milky Way. What a fascinating history we have, a story discovered by efforts of thousands of ingenious minds, for us to enjoy, to learn, and to love!

As our species left Africa and explored the other continents one by one, our extraordinarily large brains fostered our ability to feel the wonder and curiosity of it all. People have long wondered about their natural home. It is hard for us, in this age of scientific exploration, to presently comprehend the remarkable journey that our mother earth has been through, across the unfathomable eons of time. Lacking any knowledge of the earth's true dimensions or venerable age, our human ancestors told richly creative stories about their beloved homelands that supplied a sense of true place and rightful belonging.

The earthly landscape below and the bright lights above seemed like all of creation for most religions. In its infancy, humanity gazed around at its cradling lands and skies, happy to be at the center of everything to be seen. Mythological tales of creation rarely distinguished between the establishment of the whole cosmos and the origin of the earth. Our earth was surely created "in the beginning."

Young Earth or Old Earth?
The Book of Genesis in the Hebrew Old Testament can be literally read as scheduling just a handful of days between the start of nature and the formation of the earth. Vedic Hinduism assumes that the earth is about as old as everything else—as old as many trillions of years, if calculations are undertaken from the immense number of years constituting just one "day" of the long life of the primary god Brahma. All of creation was accomplished on a tightly scheduled plan, in less than one week, for Judaism. Traditional Hinduism has few details about creation, and the math gets fuzzy fast, but it can't be faulted for not trying to dramatically broaden one's sense of deep time. Buddhism similarly regards the universe and the earth as so extremely old as to be beyond human comprehension. The diverse kinds of Buddhism haven't

typically focused on assigning any specific age to the world. According to both Hinduism and Buddhism, time occurs in a cyclical manner. When the end of a great period is reached, sometimes culminating with a catastrophic destruction, then everything is reborn to start all over again. In some versions of Hinduism, we are now at the end of the Earth's last and fourth *yuga*, the *Kali Yuga*, and our world is about to come to an end, but there are assurances that it will be reborn again.

Christians reading the Bible in a literal fashion have made much shorter calculations of the age of the world. The extensive, repetitive, and frequently discordant lists of genealogies in various books of the Bible tempt Biblical literalists to trace the descendants of Adam down to Jesus. Bishop Ussher's notorious 1650 estimate calculated that creation began on 23 October 4004 BC. Neither Roman Catholicism nor any organized Protestant denomination officially sanctioned Bishop Ussher's estimate of the age of the earth based on such accounting, but some later Protestant sects enshrined as core doctrine Biblical literalism and chronologies such as Ussher's.

Most Christians today are not "young earth" creationists. Most Christians either regard the Bible as too vague about durations of time to know for sure, or they regard science as the best judge of the universe's age, although they still believe that God is responsible for the universe's origin. Similarly, Islam agrees with creationism in general, but most Muslims don't care to guess the world's age, content to leave knowledge of such cosmic matters to Allah alone. The Qur'an offers few details about a specific moment of creation or the duration of time that has passed since creation, so Islam hasn't officially decreed a timeline for the earth's age.

Religion's inability to converge on any warranted estimate about the earth's origins or its age has never stopped it from resisting science. Even in the civilizations where some early science flourished, the religious establishment regarded such speculations as innovations to be absorbed into theology or as rivals to be stamped out. Some periods of peace between science and religion have occurred, in such periods as the Hellenistic Period (circa 323–30 BCE) and Medieval Islam (circa 825–1450 AD). However, peace could be sustained only because science had to accommodate itself to the theological trends of the times. In the West, Christianity and Islam had already incorporated the much older notion that the earth was at the center of the universe. Most educated Greeks, like Plato, assumed that the earth had to be at the center of the universe, because anyone can "observe" the encircling lights in the skies. While a few Greeks like Aristarchus of Samos had figured out by 200 BC that the earth and planets encircled the sun, in the "heliocentric" model of the solar system, everyone else was working with the "geocentric model" and its geometry of circles. The mathematics of ellipses and hyperbolic curves was mostly beyond the Greeks, so they kept using the one geometric figure that allowed the easiest calculations. By keeping the earth at the center and placing the planets on a complicated system of off-center primary circles and smaller circles called epicycles rotating on those greater circles, the odd motions of the planets seemed to be predictable.

The popular religions of Hellenic times appeared to care little for whatever the epicycle astronomers were saying. However, by the time that the astronomer Ptolemy published his *Almagest* in the second century CE, early Christianity took notice. Ptolemy was still working with the geocentric epicycle model. Although the complicated mathematics of that model was still unable to account for most of the planets' motions, it remained the dominant theory. Just as Christian theologians of the second and third centuries were incorporating ideas from Plato and Aristotle in order to attain credibility among educated Mediterranean peoples, they swallowed the Ptolemaic geocentric model too. Geocentrism was conveniently consistent with Bible passages apparently denying that the earth moves while the sun does have motion, such as Joshua 10:12–13, 1 Chronicles 16:30, Psalm 104:5, and Ecclesiastes 1:5. Geocentrism also matched the religious notion that the earth had to be the most important thing in the natural world. The geocentric model of the solar system was soon elevated to the status of Church doctrine.

The notion that the earth had the shape of a flat disk, rather than a sphere, had long been a widespread belief among uneducated peoples almost everywhere. Ancient religions from Mesopotamia and Egypt to

India, China, and Japan commonly depicted a flat earth surrounded by waters. Educated Greeks had largely accepted the idea of the spherical earth, hearing the tales of sailors to southern seas who saw the northern stars lower to the horizon and new constellations rising in the southern skies. Astronomers could also notice a curved shadow of the earth cross the moon during lunar eclipses, and they noticed how the sun cast a longer shadow at a more northern location than a southern location on the same day. Most scientists, philosophers, and theologians from Aristotle (fourth century BCE) and Ptolemy (second century CE) all the way to St. Thomas Aquinas (thirteenth century AD) asserted that the earth was a sphere. A spherical earth also became acceptable to Medieval Islam. Abu Rayhan Biruni (973–1048 CE) calculated the Earth's circumference so accurately that he was only a few miles off from the current measurement of 24,901.55 miles.

It took much longer for acceptance of the idea that the earth was spinning on its north-south axis, making it appear that the heavens are revolving around the earth. A round but stationary earth at the center of the encircling solar system was enshrined by Christianity and Islam as unquestionable theology. The Polish astronomer Copernicus (1473–1543) had the mathematical skills and bold intellect to be the first significant Western scientist to publish a heliocentric theory of the solar system. Copernicus' hypothesis contradicted the Old Testament view, then enshrined as Church dogma, about the Sun's movement around the earth. His model was not as revolutionary mathematically as in its wider cultural impact. Copernicus said that the earth goes around the sun and rotates on its axis to cause day and night. Yet he still assumed that heavenly bodies trace perfectly circular motions, and his system included the same kinds of epicycles that Ptolemy had used. By ignoring other Aristotelian assumptions and proposing that the stars are much farther away than the system of sun and planets, Copernicus dismissed some key obstacles to heliocentrism.

Copernicus died before any retribution could be carried out against him from the Catholic Church. He had carefully avoided conflict with the Church because he understood the potential consequences of openly supporting his heliocentric view. Later scientists were persecuted for

GAILILEO   COPERNICUS

defending heliocentrism and anti-Aristotelian views, such as Giordano Bruno, who was burned on the stake as a heretic in 1600, and Galileo Galilei, who was forced to live under house arrest from 1633 until his death. The heliocentric model was regarded as heretical by the Roman Catholic Church for a couple more centuries, and Protestant sects were disdainful towards any contradictions with the Bible. Scientists were mostly persuaded by the mid-1600s, after Johannes Kepler showed how the planets tracing out ellipses made a much better fit with careful planetary observations, and then Isaac Newton described the laws of gravity and motion responsible for the planets held in elliptical orbits around the sun.

By the early 1700s, scientists began talking about the "solar system" as just one family of gravitationally attracted bodies, among the many other solar systems across space. The notion that the stars were suns themselves, having their own planetary systems, was advocated by Bruno but few others until the 1700s. The next question began to dawn by the mid-1700s—how might a solar system come to exist? In 1734 Emanuel Swedenborg proposed the nebular hypothesis, suggesting that the sun and the planets were formed out of a gathering and spinning cloud of small particles. Immanuel Kant elaborated this theory in 1755, adding the suggestion that observed nebula could be cradles of new systems. Pierre-Simon Laplace in turn tried to solve this theory's mathematical problems in the late 1700s. It was difficult to understand whether the sun coalesced first and only later captured (or disgorged) the planets, or perhaps the sun and the planets formed at the same time. It was not until the later twentieth century did one specific version of the nebular hypothesis, describing the sun and the planets forming out of a vast cloud of dust nearly simultaneously, receive fairly conclusive support.

Eighteenth and nineteenth century scientists had great difficulty explaining the formation of the solar system because they had few clues about the true age of the universe. They confronted questions such as, where would all that nebular dusty material come from? and why would space be so dusty? Nowadays we understand how a very old and turbulent universe would be a scene of continual stellar formation, decay, and destruction, leaving behind plenty of reusable dust and gas to coalesce over and over again. But that picture of the universe assumes what we know to be true, that the universe began many billions of years ago in an immense burst of matter. Although early scientists were liberating themselves from the Bible's cramped chronology, it was unusual for anyone to leap to the opposite thought that the world is millions or even billions of years old. Astronomy had come a long way by 1800, but it couldn't judge the universe's age until the early twentieth century. The next emerging science to help with the earth's origins was geology.

The Dynamic Earth

During the eighteenth century, the best clues about the earth's formation and age were not over peoples' heads in the skies, but under their feet in the ground. It was realized that the process of Earth's creation and evolution occurs on timescales that are unimaginable. It does not mean that things always happen incredibly slowly; in fact, some things happen quickly and catastrophically, but such events are sparsely distributed across vast expanses of time.

Today we know that our planet is 23,000 times older than our race of Homo sapiens. Several scientists have offered analogies to describe this. John McPhee asks you to consider the length of your arm as the age of the Earth and your shoulder as being the start of the Earth, the end of your finger will then be the start of the modern day. If you were to take a nail file and very lightly wipe it across the end of your fingertip, you would erase all of human civilization. That is how surprisingly short our history is compared to the Earth's.

People have long been fascinated with time. It has been a part of our philosophies and religions from their very inception. Clues that the earth had been changing for a long time had been noticed long before. Aristotle suggested that fossils of sea animals were found in the hills because the land had moved long ago. The Persian thinker Ibn Sina (also called Avicenna, eleventh century CE) added his idea that layers of rock strata first form from eroded sediments falling to the bottom of oceans to compress into rock, and these layers later get lifted back up when mountains rise up.

The infancy of geology was a time when curious naturalists literally crawled and climbed over any striking features of their surroundings. During the late 1700s, a Scottish farmer named James Hutton (1726–1797), one of the most brilliant minds of the eighteenth century, drew attention to the unusual rock formations that could be seen across the rough landscape of his native lands. He had become fixated upon the way different kinds of rocks could be observed to layer over each other in repetitive patterns, and even to interpenetrate each other, as if a harder kind of rock had pushed through softer rock when it was more fluid. Hutton's obsession was explaining why rock strata would take such complex layers and formations, and why rock strata would often be lying in tilted directions or even a vertical direction sticking up out of the ground. Evidently something powerful was responsible for pushing around immense amounts of rock, to the size of entire mountains. Hutton was the first to suppose that the Earth must be very hot deep underground. This theory, which he named Plutonism, described the earth's surface as constantly subjected, due to the erupting energies of this heat, to many slow repeating cycles of surface motion and modification.

Hutton essentially described what can be called the rock cycle: new hot rock reaching the surface to displace older rock, the weathering and erosion of rock, the ways that deposits of sediment are compressed to make rock layers, and lithification, the reformation of new rock from old. He relied on his novel idea of uniformitarianism, which refers to the way that the same basic processes slowly modifying the earth's surface can explain all of its features, so long as the Earth is actually extremely old.

In publications from 1785 until 1795, Hutton stunned the scientific world with his arguments that different kinds of rock strata had several origins, ranging from gradual accumulation of sediments at the bottom of oceans to solidification from hot lava from the depths of the planet. These hypotheses depict a constantly changing earth, with hot rock coming up from the earth's molten core and cold rock getting lifted up from the seas to the heights of continents and mountain ranges. But Hutton realized that nothing happening so fast could be observed at present, so he figured that vast amounts of time must have passed permitting very gradual processes to produce the observable results. Hutton became comfortable talking about slow events happening to the earth over millions of years, at least. But other scientists were uncomfortable with such an old earth, and they staunchly resisted Hutton's radical ideas. Although Hutton has been considered to be the father of modern geology, he did not possess persuasive writing skills. His book *An Investigation of the Principles of Knowledge and the Progress of Reason, from Sense to Science and Philosophy* was boringly pedantic at 2,138 pages in length. Few scientists or intellectuals read it, but the basic idea of uniformitarianism was immediately controversial, when it was realized that this new geology apparently contradicted the Bible.

Christians have always preferred a limited and linear way of thinking. The conflict between the cyclical and linear view of the concept of time played a substantial role in eighteenth-century geology. In Europe before Hutton, the unevenness of the land had long been explained by Christians who supposed that great catastrophes over the past few thousand years, some of them mentioned in the Bible, caused great upheavals to the earth's surface. The tale of Noah's flood in the Jewish book of Genesis was often cited as proof that an earthwide flood could dramatically rearrange the features of the earth's lands and mountains. Christianity is not alone in accepting far older myths about catastrophic events caused by gods. Myths about terrible destruction and chaotic tempests happening sometime after the original creation of the world are so common that they can be found in the myths of peoples in most every region of the world. Floods and earthquakes are especially common kinds of disasters that strike at a vulnerable time when some of the earliest people must be rescued. The way that myths mentioning a great flood are found across cultures from every region of every inhabited continent is a feature of early religion that has not yet been explained.

Early mythic cultures were evidently interested in explaining how the observable lands around them came to be the way they are. Back then, there was no way to determine the age of rocks, and no scientific method to calculate the true age of the Earth. This ignorance was bestowed upon Christian Europe, which continued to rely on the Bible. By the time amateur geology was underway, the Western intellectual world was therefore convulsed by this sharp debate over whether geological processes have a cyclical pattern of repeating itself over and over—uniformitarianism—or whether they only happen during a few cataclysmic events, starting off with a dramatic creation of everything in the beginning and concluding violently at the end—catastrophism.

One of the leaders in this conflict was Abraham Werner, who advocated a catastrophism theory called Neptunism. This theory advanced the notion that all the layers of the earth's rocks chaotically formed in a short amount of time, settling out of a world-encompassing ocean. Established churches admired this idea because it already viewed such a giant ocean as the waters mentioned in the Biblical story of Noah's flood.

As the nineteenth century unfolded, more and more geologists came to prefer the gradualist uniformitarian theory. They added another crucial piece of evidence for the slowly changing earth by proposing that continents separate, drift, and collide as they crawl inch by inch each century across the earth's surface. During the 1800s, geologists found more and more ways to see how the shape of one continent made a match with another continent, like two jigsaw puzzle pieces which had been joined but then pulled apart, and now separated by ocean waters. The close match between the eastern sides of North and South America with the western sides of Europe and Africa was an especially powerful piece of evidence. The growing idea that most of the earth consists of molten rock, permitting cold hard crust to float and drift above that core, helped the "continental drift" hypothesis.

Biology added its own observations to assist geology's idea of an extremely old Earth. Soon after Hutton's publication of his book *Theory of the Earth* in 1788, the French naturalist and paleontologist Georges

Cuvier (1769–1832) published studies of his anatomical comparisons between the skeletons of living animals and those of recently discovered fossils. The fossil remains of mammoths, for example, are different from existing species of elephants. Cuvier agreed that vast amounts of time had passed since those mammoths were alive. During the early nineteenth century naturalists were also discovering how animal species living in South America shared resemblances to species in Africa, although it seemed impossible for those species to cross the Atlantic. If both species were descended from a common ancestor who lived long ago when the two continents were very close or joined together, this commonality could be explained. Although biologists were then becoming comfortable with the idea that life has existed in various forms for very long periods of time, they were not quite ready to believe that the various species shared extinct ancestors. Cuvier to his death stubbornly rejected the very idea of evolution from common ancestors. Remaining firmly within the general Christian worldview, Cuvier affirmed the geology of catastrophism, arguing that the fossil record displays how species occasionally die off because they were all killed by sudden catastrophic events. When would biology and geology be able to agree on the earth's age?

The decisive turn for the young field of geology occurred when another Scottish geologist arrived: a great champion of Hutton and uniformitarianism, Charles Lyell (1797–1875). A brilliant writer, Lyell composed his *Principles of Geology* in 1830 and *Elements of Geology* in 1838, together establishing the foundations of modern geology. These books were filled with ideas taken from Hutton. Uniformitarianism finally won its victory with Lyell's books. He wrote "the present is key to the past," echoing Hutton, who declared, "We find no vestige of a beginning, no prospect of an end." Lyell's well-defended theories about the Earth's slow changes and great age paved the way for entire new scientific fields, such as biological evolution. Charles Darwin (1809–1882) and his 1859 *On the Origin of Species* later converted most biologists towards agreeing with a very old earth. However, Darwin himself had to admit that the bursts of new life forms and the sudden deaths of entire species, as recorded in strata of fossils, were serious challenges to his theory of long gradual evolution. That controversy was bequeathed to twentieth century biology. Geology once again made a large contribution to the debate. With the discovery that an asteroid hit the earth 65 million years ago, disrupting the climate and ending the age of dinosaurs, science incorporated some catastrophism. Both biology and geology now agree that most things happen on earth extremely slowly and continuously, occasionally punctuated by catastrophic events. These abrupt events, such as asteroid collisions, massive volcanic eruptions, or ice ages, are still entirely natural and require tens of thousands of years, at minimum, to produce their cataclysmic effects on life.

By the late 1800s most scientists disagreed with Christianity, arriving at a general consensus that the earth could not be very young. It proved to be the science of geology that finally decided the issue of exactly how old the Earth had to be. The continental drift theory remained controversial as the twentieth century began, however, because no one knew why continents would move in any particular way. In the meantime, some geological estimates of the earth's age approached and then surpassed one billion years, with some geologists suggesting that the Earth might have existed forever, but those estimates tended to suppose that Earth long ago enjoyed about the same general conditions as now. Other scientists started with a different assumption, that the Earth was presently cooling off from a much hotter molten stage long ago. That assumption was on the right track, but those nineteenth-century scientists were only working with simplistic thermodynamics. Their estimates figured that the Earth would have completely cooled in only tens of millions of years, and that the Sun would be burned out after a similar duration of time. One of England's greatest scientists, Lord Kelvin (1824–1907), calculated that the Earth wasn't older than about 40 million years old. A devout Christian, Lord Kelvin notoriously debated other geologists during the 1880s and 90s, along with Darwinian evolutionists such as T. H. Huxley who defended a much longer age of the earth to supply enough time for natural evolution.

Science often supplies some irony to its story; right when Kelvin was announcing his firmest calculations, the field of conventional thermodynamics was revolutionized. The view that the Sun and Earth must be extremely old, far older than just a few million years, received some novel support after radioactivity was discovered in 1896 by French scientist Henri Becquerel (1852–1908). Within another generation, insights by physicists into nuclear fission and fusion explained how the stars work. With nuclear fusion powering the Sun, it could remain powered for at least 10 billion years. The Earth would stay warm for that long, too. With enough radioactive elements immersed in the Earth's core—mostly potassium-40, uranium-238, uranium-235, and thorium-232—that extra radioactive heat would require many billions of years for significant cooling. Furthermore, by measuring the radioactivity of rocks from what appeared to be the oldest strata, some rocks are evidently over four billion years old. The current best measurement of the oldest rocks on Earth gives an age of about 4.5 billion years, and adding other recent evidence yields a more precise age of 4,567,000,000 years. As for the universe's age, the combination of physics and astronomy supplied the cosmological discoveries about the universe's Big Bang origins over 13 billion years ago. The earth was a very old place, located in an even older universe.

The theory of continental drift was gradually replaced by plate tectonic theory. German scientist Alfred Wegener proposed in the 1910s the new idea that our continents are the upper visible parts of immense plates of earth's crust, and the oceans also rest on more of these plates. Some of the most important changes to the earth's crust would therefore take place at the joints of these plates, where new hot magma from the core could force its way upwards, or where two moving plates would be pushed together. Most of these joints would be underwater, and little was then known about the oceans' sea beds. During World War II and the subsequent Cold War, the United States Navy mapped much of the world's ocean floors. Those plate joints were indeed discovered, in the form of deep trenches and underwater mountain ranges where plates were colliding, and long lines of volcanoes on the ocean floor where fresh upwelling lava was pushing plates apart.

The motion of continents was finally explained: continents are not randomly drifting. According to the current theory of plate tectonics, continents and the other plates of thin crust on the earth's surface get pushed around by hotter forces deeper in the earth. How hot is the earth's molten rock? By the time it reaches the earth's surface at volcanic sites, the temperature of fresh lava can range from 700° to 1200° C (1300° to 2200° F). Deeper in the earth's mantle, the bulky portion between the crust and the central core, temperatures range up to 4,000° C (7,230° F) Geophysicists have developed techniques using seismic waves from earthquakes to infer the three-dimensional structure of the earth's interior. The fluid mantle rock, slowly moving in cyclical patterns of convection from the core towards the crust and back again, is responsible for the deep forces that create plate tectonics. The mid-ocean ridge system wraps around the Earth like a snake. The slow drift of the Earth's tectonic plates, at the same rate at which human hair or nails grow—about 2–3 inches each year—splits the ridges apart, with molten rock welling up to continuously fill the ever-growing gaps. As the ocean floor moves away from the lines of new hot rock coming up, it pushes old rock around, either to jut up in the form of mountains, or to go back down into the hot core at the subduction zones, such as the Mariana trench, one of the deepest ocean floors.

This plate tectonic theory accounts for the major features of the Earth's entire surface, from mountain ranges, islands, volcanoes to earthquakes and the "Ring of Fire," the long trail of volcanic activity from Indonesia up to Alaska and back down again to the bottom of South America. Typically not faster than two inches per year, this plate movement is capable of explaining vast and dramatic changes to the earth's surface in just tens of millions of years. Over the passage of billions of years, the entire surface of the Earth has been renewed, eroded, and recycled many times. Earthquakes, volcanic eruptions, and tsunamis are the byproducts of our planet's ceaseless activity. Volcanic activity has also produced most of the earth's atmosphere.

It required the combined efforts of several emerging sciences, including astronomy, geology, paleontology, biology, chemistry, mineralogy, hydrology, atmospheric science, thermodynamics, and nuclear physics, to assemble a convincing explanation of why the earth must be extremely old. Many well-established sciences would have to be mostly wrong (and they are not) to make any room for a belief that we inhabit a young earth.

The Earth's Cradle

The real story of Earth and its dramatic changes since its birth should amaze us. The secrets of our beautiful planet, our cool blue-green oasis, ultimately are about a fiery birth from a cold grave. The leftover dust and drifting debris from many past solar systems, going all the way back to our galaxy's own origins 13 billion years ago, didn't just evaporate away. Gravity always pulls back whatever happens to fall apart. That nebula of gas, dust, and meteorites pulled together under its own gravitational attraction. The way gravity works on clouds of particles means that it gradually introduces a swirling effect as the nebula contracts together and the nebulas' center becomes much denser than the rest. Over time, a spinning disk of particles, larger than the current solar system, grouped itself into distinct bulges as smaller swirls of condensing matter formed within the larger whole. This nebular portrait of the beginnings of our solar system has been seen happening elsewhere in our galaxy by powerful telescopes. When the central ball of matter become dense enough to ignite nuclear fusion, the baby star is born and lights up the rest of the spinning nebula.

Our solar system's star, the Sun, illuminated a new solar system. But the births of the planets took more time as they each accumulated the remaining particulate matter around them. That early sun shone upon many thousands of brand new planetoids busily colliding into each other. The planetoid that became our earth kept growing in size as all that material rained down and hit the surface through a process called accretion. The molten sphere of rock that was the baby earth did not have a chance to cool off for hundreds of millions of years as everything from the size of small rocks to huge continent-sized planetoids smashed into it at thousands of miles per hour. This era has been subdivided into early and late bombardment periods in this specialized field of study, called planetary science. Planetary science is closely involved with the effort to explain why the earth has such a large single moon. According to one theory, the moon was formed when some large planetoid hit the earth during early bombardment and the collision rebounded, with a large chunk broken off the earth to form the moon. Another theory suggests that two smaller planetoids got close to the earth only to merge together instead of merging with the earth in order to form the moon. The early solar system was indeed a chaotic place. Most of that early debris is now gone, swallowed up by the sun or the planets, or thrown entirely out of the solar system. Many asteroids within the solar system and comets arriving from beyond Pluto still wander through the solar system. A few meteorites reaching the earth have been dated to ages slightly older than the earth, indicating how the solar system is still hit with matter from that original nebula.

Approximately seven or eight hundred million years after the Earth's beginnings, the surface had cooled enough to permit a thin layer of crust to supply solid ground. This thin crust, only a few miles deep in some places, contains most of the lighter kinds of rock and interesting metals like tin, copper, and aluminum. The reason why the earth's crust is so much lighter compared to the deeper layers is because denser materials always sink deeper into the core while less dense materials rise to the surface. The lightest molecules, such as water and carbon dioxide, made their way to the surface quickly. The steamy waters that had bubbled up soon cooled and rained back down for millions and millions of years to make the first oceans. The entire planet was soon a water world, and it was destined to remain so. The oceans soon became hospitable for the origin of primitive life, perhaps by the time that the earth was just a billion years old, or even earlier. Leftover material from the original nebula continued to occasionally hit the Earth, as smaller meteors and bigger asteroids and comets crashed down. Some of these catastrophic impacts altered the earth's climate and contributed to massive deaths of thousands of species. For example, the dinosaurs did not survive the colder climate caused by an asteroid's collision with the earth 65 million years ago. But life is extremely robust, as other species evolved to take advantage of changing environments. Life as a whole tends to regard destructive catastrophes as creative opportunities. From death, comes life. We now fear big collisions with asteroids and comets.

But the Earth could not be such a richly productive planet if those collisions hadn't happened.

All of the important elements that make up the Earth's complex surface, the elements combining for mineral deposits and organic compounds, were swept up from that original nebula as the Earth grew. That original nebula contained all those elements because it consisted of the remnants of many solar systems, long dead and destroyed. Stars and their planetary systems had been forming and falling apart in the Milky Way for eight billion years before our solar system pulled together. Some of the Earth's basic elements—hydrogen, helium, and lithium—were created out of the Big Bang. The rest of the elements were forged in the hot furnaces of stars that had lived and died before our own Sun. Many stars died for us so we could live.

Nuclear fusion in the core of stars welds lighter elements like hydrogen and helium together to make the bigger elements of carbon, oxygen, and nitrogen, and further fusion welds those elements to make heavier elements such as magnesium, sulfur, and iron. The dramatic death of a large star which burns brightly but burns out faster, climaxes in an immense explosion called a supernova. When a very old star cannot sustain ordinary nuclear fusion at its core past the element of iron, the core cools and loses its ability to keep the rest of the star inflated. The entire star quickly collapses inward, sending its temperature soaring so high that heavier elements than iron are formed, but this collapse spells doom for the star. The immense pressures generated at its core causes an explosion to destroy the star in less than a second. That explosion forges the star's smaller atoms into dozens of the heaviest elements, all the way to uranium. That is why elements heavier than iron are found in such miniscule quantities in our universe. Other kinds of supernovae, called Type IA supernovae, involve a binary system of two stars collapsing into each other to cause the explosion of a white dwarf once it reaches 1.4 times the size of our sun. Interestingly, Type IA supernovae have been extremely useful, because they allowed astronomers to figure out that our universe was expanding and accelerating. These "standard candles" permit the measurement of cosmic distances more accurately, because they always explode with the same brightness and energy no matter where they are located in the universe, and thus they allow us to use a scale of comparison of brightness with distance. Some of the biggest stars, on the other hand, end up as black holes, too dense and gravitationally powerful at their cores to permit any more light to escape.

After stars die, whether from explosive supernovas or from smaller novas that leave behind cool dwarf suns or white neutron stars, they leave behind the surviving wreckage of their planetary systems. These small stellar remnants are hard enough to spot by powerful telescopes; interstellar drifting planets and asteroids are so much harder to detect. Astronomers are becoming confident that our galaxy is completely littered with this sort of leftover debris. Unmoored from their orbits, or thrown completely out of orbit, these rocky planets and planetoids are left behind to wandering between the still-burning stars. Perhaps hundreds of billions of orphaned planets are scattered throughout our galaxy. Some of that drifting debris ended up in our system's original nebula. And stars are always making more. How many stars had to die in order to forge the many dozens of elements crucial for a new planet like Earth to enjoy a complex surface and sustain an even more complex environment for life? Perhaps many dozens of stars died for the Earth. Once again, we see how from death, comes life.

The Sun, the Earth, and the Moon

The Earth has a truly special relationship with the Sun. Its distance from the Sun, and its placement in the order of planets outward from the Sun to Neptune, has a critical significance for the way the Earth has evolved. Unlike other typical kinds of solar systems, with multiple stars dancing around each other or huge gas giant planets sweeping close in to their suns, our solar system displays a more orderly pattern. With just one star in our solar system, and the gas giants—Jupiter, Saturn, Uranus, and Neptune—forming farther out from the Sun, the closer and smaller rocky planets of Mercury, Venus, Earth, and Mars have had a chance to stay in fairly stable orbits since their formation. Stable orbits can be easy targets for the renegade asteroids and comets to eventually hit, and many have hit the inner planets, including the

Earth. However, after the Earth's formation, those big collisions have been less and less frequent, thanks to the four gas giants patrolling the outer reaches of the solar system. Much of the leftover debris from the original nebula was swallowed up in the gaseous embrace of their larger gravitational attraction. Almost everything that didn't get absorbed got flung around by the gas giants and sent back out into the region of planetoid and asteroid debris beyond Neptune, called the Kuiper Belt, or thrown even as far as the Oort Cloud that has only the slightest gravitational tie to the Sun.

Not too close to the Sun, permitting liquid water to slosh across the Earth's crust, and not to close to the powerful but protective gas giants, the Earth has long occupied a "habitable zone" of the solar system. This "habitable zone" only means habitable for life like us—we need lots of liquid water lying on the planetary crust, with only a little water steam and ice around. Other kinds of life could take advantage of different conditions. Perhaps enough liquid water to sustain life is buried deep under the frigid surface ice of some moons of Jupiter or Saturn. Perhaps liquid methane, also common enough on the larger moons like Saturn's Titan, permits a much colder kind of life. More exploration of our solar system, and perhaps exploration of other solar systems, will discover how different sorts of "habitable zones" each provide their own unique conditions for unique kinds of life forms. Just as human explorers had to go see for themselves how life could flourish on the heights of mountains and depths of oceans, future explorers will discover where life flourishes in all kinds of niches within solar systems. From the little that we now comprehend about the chemistry of life in general, it appears that life probably sticks close to stars.

Common wisdom long whispered to primitive humanity that the Sun is the ultimate power for our life. A catalog of the mythologies regarding the Sun as a god, or identifying the Sun as a manifestation of a god's powers, would be a list of almost every ancient religion and many contemporary religions. A sort of simplistic naturalism animated humanity's first religions, and their nature-worshipping mythologies found the gods among the powerful and important forces of nature. The names of Sun gods may sound unfamiliar or strange, but each one was worshipped by hundreds of thousands or even millions of people for centuries. The Egyptian gods Amun and Ra, the Greek god Apollo, the Roman god Sol (also a Norse name), the Persian god Mithras, the Vedic Hindu gods Mitra and Varuna (later merged into the god Surya), the Japanese goddess Amaterasu, and the Aztec gods Tonatiuh and Huitzilopochtli were all associated with the Sun to serve as powerful gods in their respective pantheons.

Over the past three thousand years, both science and religion have helped to evolve many cultures away from nature-worship. Younger theological religions, such as Christianity, Islam, and later Hinduism, made enough compromises with early sciences such as astronomy to realize that the gods were not within nature, but somehow beyond nature. As the world began to look more and more like a naturally evolving system needing no driving spirits, the gods retreated. Some impersonal divinities like Aristotle's Prime Mover, the Hindu Brahma, or the Chinese Tao receded into a mysterious energy underlying all the evident forces of nature. Other personal divinities such as the Judeo-Christian Yahweh and Islam's Allah had to be divorced from nature entirely so that some immaterial and transcendent realm could house them.

Nature religions looked at the Sun and saw an awesome God, while theological religions looked at the Sun and saw God's awesome handi-work. Looking at the Sun too much can cause blindness. Regarding the Sun as a divinity or as a divine tool blinded humanity for a very long time. Many people are still cognitively blinded. Religious people still think that it makes sense to demand special answers to questions such as "How did the Earth get made just right for life?" or "Why would the Earth have just the right distance from the Sun?" The questions assume that there is something objectively "just right" about our solar system or our planet. Naturally, we think that the Sun and the Earth works fine, more or less, for us. If we really knew that the Sun was the only sun, and the Earth was the only planet, they would indeed be special. But we now know the exact opposite: trillions of suns, and many trillions of planets and planetoids populate the many billions of galaxies strewn across the universe. They have innumerable combinations of varieties of solar systems, some with gas giants near and rocky planets far away; others with only gas giants; and yet still others with random orders of gas giants, asteroid belts, and rocky planets, and too many other varieties to list. Using the Kepler space telescope, astronomers have detected thousands of "exoplanets" orbiting stars in our galactic neighborhood. Among that number, over one hundred planets appear to be situated within a "habitable zone" permitting liquid water on the surface, just as the Earth occupies such a favorable distance from its Sun for supporting life like ours. Some of those planets are close to Earth's size and receive about as much light from their stars as the Earth does from the Sun. The idea that there are planets similar to Earth has received important scientific confirmations.

Is the Earth all that special, from a cosmic perspective? Earth does have it pretty good. It may turn out that rocky planets having stable orbits within a liquid-water habitable zone and protected by gas giants are in a small minority among all those other planetary combinations. But a small percentage of an immense number is still a very large number. If only one in a million solar systems has a configuration closely resembling ours, there would be hundreds of millions of planets similar to Earth in the universe, and as many as a hundred thousand "earths" in our galaxy alone. A galaxy is a roomy place, so a galactic explorer traveling much faster than light would infrequently encounter one of these sister earths. However, the nearest planet quite similar to Earth could be just dozens of light years away from us. Also, many rocky wet planets ranging from the size of six earths to the size of Mercury are probably sprinkled around a region of 100 light years' distance. Some planets around that size have already come into our range of detection, and many more will be discovered by the next generation of space telescopes. A planet very similar to Earth (only somewhat larger) that has plenty of water has already been detected. In coming decades, a planet quite similar to Earth will be discovered, although the presence of life on such a planet may remain a curiosity.

Our solar system permitted the origin and evolution of our kind of life. We are here because the Sun and the Earth were here first. Obviously, we regard that situation as pretty special, and we should

regard it as special to us, since we would have to travel great distances to find another planet just like our Earth. However, given so many kinds of solar systems with their own assortments of planets, the objective "odds" that a few planets in our neighborhood of the galaxy are similar to the Earth must be very good. It was inevitable that many earthlike planets would be formed among all the kinds of planets possible, and we happened to evolve on one of them. The fact that for most of our species' existence we only knew about one earth explains why religions demand and supply answers about the Earth's "special" nature. But the time for supposing that our Earth is truly rare or special, as far as the whole universe is concerned, has now passed entirely. The Earth may be a peculiar planet, but it isn't special or important in any objective way. Questions about why the Earth is so special only exposes the questioner's own ignorance. Offering religious "answers" to such ignorant "questions" is an even deeper kind of anti-intellectual game, frequently played by theologians, often with little or no instruction in the methods of science. Our Earth was never designed for us; our species has evolved and managed to survive here, at least for now. But our fate could be similar to that of the dinosaurs, sent to extinction by some planetary catastrophe or cosmic accident.

The Sun was the bringer of life on earth, and it will be the bringer of its death. The Sun has long showered the Earth with enough light radiation to provide plenty of energy for the fast chemistry of organic life. When the Sun dies, any life still on Earth will die. About five billion years from now, with its core fusion processes exhausted, the Sun will bloat up into a red giant, and its surface will reach Earth's current orbit. It will either get vaporized by the Sun, or barely survive the Sun's red giant phase as a hot molten ball, much like what it was at its birth. After the outer layers of this red giant Sun blows away into the outer reaches of the solar system, anything that has survived will slowly drift away from what is left of the Sun. At the center of this mess there will just be a hot white dwarf made mostly of carbon and oxygen about the size of the Earth.

The Sun's death will leave behind a graveyard, a scattered scene of rocky devastation where clouds of particulate minerals and chunks of planetoids and asteroids are all that is left. But nature never really leaves behind a wasteland—graveyards are also fertile fields. Everything is always mixed up and recycled by gravity and time. Some new sun will grow from that nebular chaos and shine upon a fresh arrangement of planets, perhaps fifteen billion years from now, where our solar system used to be. The elements of life on Earth might get reused for new life on some lucky future planet yet to born.

After the Sun, the other large body in the sky, the Moon, has long attracted humanity's attention. Ancient astronomy tried to track many heavenly objects, but it had the most success determining the length of the solar year, the timing of the periodic lunar cycles, and the predictions of solar and lunar eclipses. Babylonian, Indian, and Chinese astronomers before the time of Ptolemy had track records of success at predicting eclipses. Greek thinkers, such as Anaxagoras (fourth century BCE), proposed that both the Sun and the Moon were giant spherical rocks like the Earth. Some Greek and Chinese astronomers, along with later Medieval Islamic and Indian astronomers, added that the Moon can only reflect light from the Sun. Ptolemy himself made a fairly accurate calculation of the Moon's distance from the Earth, which is now measured at about 239,000 miles, averaged. Other early thinkers figured out that the Moon is responsible for the ocean tides on Earth.

The early Christians fell behind other cultures because of their attachment to Aristotelian notions that the Moon transcended the natural realm of earthly matter, and to certain Biblical passages about the Sun and Moon. For example, Isaiah 30:26 seems to say that both the Sun and the Moon shine their own light. Until the time of Galileo, most Christians supposed that the Moon and other heavenly bodies must be perfectly shaped without any deformities, right along with the perfectly circular paths they were supposed to follow. When Galileo pointed his improved telescope at the Moon, he perceived its surface irregularities immediately. He described and depicted what he saw in his 1610 book *Sidereus Nuncius* (Sidereal Messenger). The Moon's surface was evidently covered with brighter mountains, darker plains, and shadowed craters. Galileo also thought he saw volcanoes on the Moon, although this was an error; the Moon has been hit with many asteroids and comets, but it has not had an actively tectonic or volcanic surface since early in its history.

The Moon is an unusual companion for a rocky planet like Earth. The gas giants Jupiter and Saturn each have larger moons, but our Moon ranks fifth in size after them. The Moon's size is one quarter of the Earth's diameter, and its ratio of mass to the Earth, at 1/81, are uniquely large numbers in the solar system. The Moon has a small hot core while most of its mass consists of crusty solid rock very similar to the Earth's own upper crust. Its smaller size and tiny core means that it cooled off much faster and has stayed pretty frozen since. Its surface features haven't changed much since its formation, besides the magma plains and many craters due to impacts. The current plausible hypothesis for the Moon's proximity and similarity to Earth is that the

Moon was blasted off the surface of the Earth when some Mars-sized planetoid collided with the young Earth about 30 to 50 million years into its existence.

Some further odd features of the Earth-Moon relationship require explanation. As everyone can notice, we can only see one side of the Moon, and the apparent size of the Moon is almost exactly the same size as the disk of the Sun. The first fact is no coincidence, but the second fact is. In a gravitational system of two bodies, where one is much smaller (but not tiny) compared to the other, the mathematics of the gravitational, torque, and other tidal forces shows how the smaller body would gradually lock into a rotational period equal to the time it takes to make one orbit. Other effects of the Earth-Moon gravitational system include the way that the Moon used to be much closer to the Earth and is gradually moving away, and the way that the Earth's own rotational speed is gradually slowing down, making one day longer now than it was long ago. By adding into such calculations the considerable effects of the Sun's gravity on the Earth-Moon system, the Moon's motions and the oceanic tides on Earth have been thoroughly explained. We just happen to be observing the skies when the Moon can entirely hide the Sun in an eclipse; hundreds of millions of years from now, an earthly observer would notice how the more distant Moon can't do that anymore.

Another important feature of the Moon's regulation of earthly matters is the way that the Moon is responsible for keeping the earth's tilt on its axis quite moderate. As the Earth wobbles while it spins, its axial tilt is about 23.4 degrees, and so the northern hemisphere leans towards the Sun during half of the year, and leans away from the Sun the other half of the year. That is exactly reversed, of course, for the southern hemisphere, which gets winter while the northern hemisphere gets summer. By contrast, the axial tilt of Venus is 177 degrees, and the other planets have their own diverse tilts. Enjoying a moderate tilt implies that the Earth's climate cannot get too excessive. There are other lunar benefits. The carving of the Moon out of the Earth mixed up the mineral composition of the young Earth's crust, and a nearby Moon pulling on the Earth's crusty surface has caused serious continental tectonics. Added together, these disruptions meant that many kinds of elements and metals stayed available near or at the Earth's surface. Without a Moon, the Earth may have been a much less habitable, interesting, and useful place.

Science has confirmed some intriguing ways that the Moon has long made a significant difference to the Earth's evolution and possibly life's evolution as well. Some notions about the Moon's influence have no basis in reality. The Moon (Luna in Latin) is not responsible for causing mental illnesses, so the old-fashioned labels of "lunacy" and "lunatic" are misleading now. Slightly more suicides occur during new moons, but that is probably because the human body and its psychological emotions can be sensitive to seasonal changes and the amount of ambient light. The similarity between the phases of the moon and the periodic timing of women's menstrual cycles (a scientific explanation awaits) has linked the Moon to female goddesses. The placement of a bright moon over a romantic scene is nowadays just an artistic element. As a lure for human exploration, perhaps nothing in the sky has served as so powerful an attraction to writers of science fiction and to hopeful astronauts. Humanity may someday step onto many more worlds, but the tale of our first journey to our own Moon will forever be part of humanity's narrative.

The Oceans and Atmosphere

Science has learned many secrets about our home planet. We now know that our beautiful Earth has been through an epic journey of both cataclysmic and gradual changes. The result is an environment that is about two-thirds water, one-third land, and a home to many millions of species of microbes, plants, and animals for over three billion years. The Earth has been a fireball, a water ball, and an ice ball; it has been a scene of fiery crashes, huge volcanoes, colliding continents, raging oceans and tsunamis, and poisonous skies. Through it all, the slow effects of steam, water, and ice have perhaps the greatest impact on the details of the Earth's surface. Whatever the land can raise up, water can erode down.

The Earth has always been home to water from its beginnings. Tiny amounts of water have been discovered in meteorites. That water has the same ratios of hydrogen isotopes as earth's water, indicating a common source. That source was the original nebula from which the Earth formed. The simple combination of two atoms of hydrogen with one atom of oxygen was quite common since the first stars made oxygen in their furnaces and then exploded with so many other gases and elements. In the original nebula that became our solar system, water (H_2O) was plentiful in the form of various isotopes, along with other simple molecular compounds such as nitrogen (N_2), ammonia (NH_3), and carbon dioxide (CO_2). There is also good evidence of water on most of the other planets and many of their moons, including small amounts on our own Moon.

Much of the Earth's water is still locked deep in the crust, although heat and pressure long ago forced most water up to the Earth's surface to form its lakes, seas, and oceans. In the form of water vapor, water is the tenth most abundant component of the atmosphere. About 71% of the Earth's surface is covered by liquid or frozen water. About 97.5% of that water is saline, while the remaining 2.5% is fresh water. About two-thirds of the fresh water is currently ice. Less than 1.8% of the surface water on the Earth is easily drinkable by animals.

Fortunately for most life on Earth, salty water is enough. Most of that saltiness comes from sodium chloride, although other salt compounds such as magnesium sulfates, calcium sulfates, and bicarbonates are washed down into the oceans to give salt water its characteristic flavor. Life originated in salt water, perhaps deep underwater near hot magma vents, and it flourished in shallow seas or near ocean shores. Life continues to use salt water; that is why your own blood tastes salty. Animals constantly lose water to evaporation and excretion. Land animals often need access to fresh water only because their food already provides plenty of salts but not enough water. Some animals need to especially refresh their bloodstream balance of salts with diet supplements by instinctively seeking out salt sources. Human athletes have to become familiar with ways of replenishing their salts as well.

Salty water has dominated the surface since its beginnings, although ocean sizes and levels have changed enormously during the Earth's long existence. Because of the continental movement caused by plate tectonics, the Earth has sometimes had all of its dry land bunched together, surrounded by a single vast ocean. This has happened at least four times. Alfred Wegener himself proposed that one "supercontinent" had long ago split apart to make today's seven continents. This most recent supercontinent has been called Pangaea, and it began to break apart around 175 million years ago. As the separate continents began to split apart, the single great ocean surrounding Pangaea, called Panthalassa, was divided into the separate oceans seen today. In the future, perhaps some 250 million years from now, a future Pangaea will reform as the Atlantic Ocean shrinks and disappears, Africa entirely merges with Europe, and Antarctica and Australia relink with southern Asia.

The other major changes to the world's oceans have to do with their levels. Presently, during a relatively warm age with small ice caps, the ocean levels are moderately high. In the past, ocean levels have been much higher during very warm climatic periods, and sometimes ocean levels have been far lower when much water was locked into snow and ice during cold climatic periods.

These two kinds of major oceanic features, the oceans' dimensions and their depths, are interrelated, and both are intricately connected with the evolution of life on Earth. For example, about 1.2 billion years ago, the supercontinent previous to Pangaea (called Rodinia) was surrounded by a single vast ocean. The wild climate during the first and second billion years of the Earth—when volcanoes were plentiful, the Moon was still close enough to the Earth to cause 500-foot ocean tides, a heavily carbon-dioxide atmosphere weighed down, and hurricane-force winds swept the lands—was then subsiding for Rodinia. Still, Rodinia was a rocky wasteland devoid of life. The tiny organisms then living in the water could not deal with an oxygenated atmosphere, or the direct ultraviolet radiation from the sun unblocked by a protective ozone layer, which hadn't accumulated. However, the formation of Rodinia and a single giant sea may have contributed, along with other atmospheric cooling processes, to a dramatic cooling of the Earth, perhaps a 40 degree temperature drop at the equator. Most land was covered with ice and snow, the ocean levels dropped dramatically, and much oceanic life was killed off. These kinds of cold durations are very disruptive to watery life, since most complex multicellular organisms were surviving near ocean coastlines and in shallow seas.

This "snowball earth" scenario had happened before, around 2.5 to 2.2 billion years ago. Much of the Earth's varieties of life were harmed by this cold event, combined with the sudden lethal oxygenation of the atmosphere by photosynthesizing cyanobacteria. They formed vast mats of colonies, called stromatolites, the first structures of life on the planet. There seems to be more than a coincidence between the formation of supercontinents and mass extinctions of life. Similarly, the breaking apart of a supercontinent can supply the favorable conditions for a warming period and plenty of shallow gentle seas. When Rodinia broke apart and continents were again far apart by 700 million years ago, life surged in response. By 580 million years ago, another supercontinent called Pannotia was split apart by fresh volcanic ridges, causing a warming to the atmosphere, and it divided into continents again. The first large and complex multicellular organisms were flourishing, the rate of evolution accelerated dramatically, and the great "Cambrian explosion" of the major phyla of life began 530 million years ago. It was a good thing that life dramatically expanded and diversified during the Cambrian explosion. By 440 million years ago, the next supercontinent Pangaea had formed, the Earth grew colder and sea levels dropped, and over 50% of all species on the planet died in another mass extinction. But the breakup of Pangaea around 170 million years ago helped to make dinosaurs the dominant species on the planet.

Fluctuations in the atmosphere are always contributing causes to planetary temperatures, ocean levels, and whether life can flourish or not. Atmospheric conditions can change due to the compositions of elements in the air, the penetration of cosmic radiation down to the surface, the spread of dust from comet collisions or volcano eruptions, and other variables. The Sun periodically drifts near the upper and lower edges of the galaxy as it slowly revolves around the galactic center along with all the neighboring stars, probably exposing the Earth to greater intergalactic radiation that can kill life or accelerate mutation. After enough oxygen accumulated in the atmosphere, new oxygen-breathing animals took advantage of those conditions, helpfully protected by the new ozone layer in the upper atmosphere blocking much radiation. Further mass extinctions occurring around 370, 250, 200, and 65 million years ago had major atmospheric changes as primary causes.

Resilient Life

Through all of the slow and fast climatic changes that the evolving Earth has endured, life has survived and even flourished. There are organisms at the tops of high mountains and at the bottom of the deepest submarine trenches. Life has penetrated into virtually every environmental niche possible, and even survives miles underground. The Earth's surface is nearly completely full of life. To trace where all of life lives, where this entire "biosphere" is located, would amount to surveying all of the earth's surface from bedrock miles down to clouds floating miles high. We literally reside within a vast realm of life. Humanity gradually came to an awareness of how our earthly surroundings are pulsing with the endless cycles of life, death, and rebirth.

Applying the all-too-human notion that the origin of life is naturally female, motherhood was long associated with the natural Earth itself. As the world's earliest prehistoric religions emerged tens of thousands of years ago, the widespread myth of Mother Earth or of the Earth Goddess easily followed, and many religions retained that idea for a long time. Was there a mothering spirit nurturing the primordial stew of organic compounds on the early earth?

Western culture has inherited her Greek and Roman manifestations. The ancient Greeks called her Gaia, while the Romans used the name Terra. Most other religions have had their equivalents. Rare exceptions include more recent theological religions that became monotheistic by selecting a male deity as the one real god. It was not easy to eliminate such a longstanding attribution of the feminine to the divinely natural. Even in early Christianity, for example, the feminine side of divinity was submerged and disguised as demonic forces, and recasting nature as corrupt and evil took great effort. Other historical religions typically retained a feminine deity as supreme, or paired her with a male god, so children could account for other specific features of the world and humanity. Ignorant of how life really works, numerous mythologies spun endless imaginative narratives about this primordial mother or goddess of nature. Personifying an entire planet, or attributing some kind of "spirit" to the Earth, has no place in any modern scientific worldview.

Viewing the Earth itself as some sort of living entity may enjoy scientific merit. What really is this thing called Life? Life is fundamentally just one particular form of thermodynamic chemistry, among the many other kinds of chemical transformations constantly cycling throughout the Earth's very active surface. Previous sections of this chapter have recounted the dynamically interlocked processes happening in the upper crust, the oceans and the land, and the atmosphere. Furthermore, the organic chemistry of life is ecologically interconnected with many other continual chemical cycles. In fact, life is so thoroughly interfused with those other chemical cycles that it is somewhat arbitrary to fix a sharp line dividing living processes from nonliving processes. The cycles of organic chemistry require an energy source to create energetic chemical bonds in compound molecules, such as the carbohydrates that animals and plants use. Plants make those energy-filled compounds themselves in photosynthesis, and photosynthesis in turn requires the light source (the sun) along with available carbon dioxide and other nutrients in the soil. Animals then consume the plants, either directly like herbivores or indirectly as carnivores eating the plant eaters. But all life would have exhausted the chemical resources laying around on the Earth's surface long ago and promptly died out. That didn't happen, because of all of the other chemical recycling processes simultaneously happening. Understanding the Earth's ecology permits us to see how life can sustain its organic cycles only because many other chemical cycles endlessly do their work, too.

From the widest perspective of thermodynamics and geochemistry, life is just one phase of the bigger process by which the Earth must shed excess energy into outer space. All of the surface thermodynamic processes can be treated most generally as a unified system which transforms inputs of available energy into dissipated forms having higher entropy. Heat from the Earth's molten core and heating from the Sun's radiation supply the input energy for everything else that happens at the Earth's surface. Without these two sources of energy, the Earth would have cooled off to reach thermodynamic equilibrium long ago. However, life and all other chemical processes are active on the Earth's surface because it cannot yet reach thermodynamic equilibrium. Kept far from thermodynamic equilibrium by those two powerful energy sources, the hot core and the hotter Sun, the Earth's surface is thermodynamically driven to redistribute, transform, and radiate that input of energy. Life is just one of the many ways that the Earth's surface is compliantly obeying the fundamental laws of thermodynamics and chemistry. Life is no accidental or peripheral phenomenon on this planet—life is an essential part of the Earth's own evolution. Without life, the Earth would have been a dramatically different place.

We ask again, what exactly is this thing called Life? Organic life is basically one phase of the Earth's vast carbon cycle, which in turn is an essential chemical component of the Earth's futile quest for thermodynamic equilibrium. Organic life by definition is based on carbon and all of the molecular compounds that can be made from carbon, when carbon bonds with other common elements such as hydrogen, oxygen, nitrogen, sulfur, phosphorus, magnesium, and potassium. These eight elements together make up more than 99% of the mass of living cells. Life also uses organic compounds to make additional bonds with compounds made from other common elements such as silicon, calcium, and iron. For example, iron is responsible for the ability of hemoglobin in animal blood to transport oxygen for the oxidation processes happening in all cells. Plants use a compound similar to hemoglobin except that magnesium is the core element rather than iron, so that hemoglobin looks red while chlorophyll looks green. Ultimately, the success of all of life's processes depends on the endless recycling of carbon. Life tears apart the chemical bonds among carbon and other elements, releasing the stored energy of those tight bonds. In the end, carbon stays bonded to a couple of other elements, especially hydrogen, oxygen, and nitrogen, as it is excreted and discarded by life. Those simplest carbon compounds just go back into the earth or the oceans as dead organic matter, or into the atmosphere as carbon dioxide, but they never rest.

The carbon cycle is the transfer of carbon between major reservoirs of carbon in the earth's crust, the oceans, the biosphere (all of organic life), and the atmosphere. If any one of these reservoirs were permitted to retain its accumulated carbon, there would soon not be enough carbon for life to continue. For example, if life itself soaked up the carbon and it remained locked within the mass of living or dead organic matter, later on there could be no fresh intake of carbon for life's processes. Even the fungi and bacteria that live on decaying organic matter would themselves die out eventually, since less and less carbohydrate energy is available with each generation. All of life and organic chemistry would eventually stop. But organic matter and its carbon never stop moving. For another example, the oceans do absorb most of the earth's available surface carbon, but the ocean waters constantly engage in complex exchanges of dissolved carbon compounds with the life, mostly phytoplankton, in the water.

These exchanges between carbon reservoirs are thermodynamically inefficient and carbon-wasteful, so fresh inputs of both heat energy and fresh carbon have played a crucial role in sustaining life. Heat from the core and the Sun has already been mentioned. As for carbon, the carbon released by volcanic activity has been a significant supplement over the long history of life on Earth. Dramatic changes to the percentage of carbon compounds in the atmosphere have been especially influential on life on the land. More or less CO_2 in the atmosphere dictates the flourishing of plant life, for example. Plants directly need that CO_2 in photosynthesis, and plants flourish when higher concentrations of carbon compounds keep the atmosphere conveniently warmer. These changing conditions naturally affect animal life on the land as well, since animal life eats the abundant plant life and breathes the oxygen released by plants. Other kinds of chemical exchanges between the carbon reservoirs, too many to catalog here, are also critical for making sure that organic chemistry remains an essential component of the wider thermodynamic processes happening on the earth's surface.

One thousand scientists at the European Geophysical Union meeting signed a declaration at its 2001 Amsterdam meeting, agreeing that "[t]he Earth System behaves as a single, self-regulating system with physical, chemical, biological, and human components." That's true enough, as far as science goes, but we still want to ask the question, is the Earth alive?

Because life is not independent or self-sufficient, and the chemical cycles essential to life proceed through oceanic, earthly, and atmospheric phases, it could be said that the whole thermodynamic system of the Earth's surface is alive. That planetary thermodynamic system is not alive in the narrow biological sense by consisting of cells with DNA, of course. However, from the objective standpoint of geochemistry, able to regard the interlocking systems of inorganic and organic chemistry as thoroughly interfused and interdependent, it is no mistake to view the earth's surface as alive. Central features of life, including the abilities to sustain and regulate its own internal structures, by maintaining constant exchanges of energy with its surroundings, and even to transform its surroundings to keep them more amenable to future use, are all displayed there. Besides these four basic features of metabolism, self-regulation, internal regeneration, and external transformation, even more features of life can be seen, such as evolution as a whole and reproductive capacity as a whole. Life as a whole has certainly evolved on earth. More than 99% of the species that have ever lived on Earth are extinct, but they have left behind quite different descendants. Life on earth could, theoretically, be partially reproduced elsewhere on some other planet if the interstellar distances could be crossed. These six traits—metabolism, self-regulation, internal regeneration, external transformation, evolution, and reproduction—are already regarded as the primary characteristics of life. We would look for exactly these six dynamic processes on other planets if we were seeking life but we could not anticipate what concrete form it might take.

During the second half of the twentieth century, the intellectual synthesis of geophysics, biochemistry, systems ecology, and related environmental fields resulted in pioneering theories that regarded the Earth's surface from bedrock up to cloud as one general system for life. This single immense thermodynamic biosphere was labeled as "Gaia" by planetary scientist and ecologist James Lovelock in the 1970s, who emphasized how Gaia was a homeostatically self-regulating and self-repairing system. He wrote that Gaia is "a complex entity involving the Earth's biosphere, atmosphere, oceans, and soil; the totality constituting a feedback or cybernetic system which seeks an optimal physical and chemical environment for life on this planet." Other scientists, such as microbiologist Lynn Margulis, gradually endorsed this Gaia theory, and together they found more and more scientific evidence that Gaia can be treated as an entity displaying those six central features of systemic life as a whole.

Three exaggerations of this Gaia theory promptly followed its entry into the scientific world. Enthusiasts about this Gaia worldview have been heard to claim that Gaia is itself a living organism, or that Gaia has some animal or even humanlike traits, or even that Gaia has spiritual or divine qualities. None of these exaggerations can receive scientific support.

Some environmentalists have hastily spoken of Gaia as a unitary organism in its own right. This is a matter that the science of biology alone can settle. Biology has its set criteria for treating something as an individual organism, and Gaia as a whole does not entirely match those criteria. Biology is not just about individual organisms, of course, having become comfortable with the study of colonies of organisms and systems of interdependent organisms. The outdated notion that a genuine organism can live and reproduce by itself only really applied to some species. Complex life is multicellular, and most multicellular organisms cooperate to some degree in groups, within species or symbiotically between species. Some species, such as certain ant species, develop into what have been called "superorganisms." A superorganism lacks physical connections among their organic parts. Ants, for example, can walk about separately—yet an entire ant colony displays behavioral traits permitting this superorganism to be alive as a whole in ways that none of its component organisms can. Individual ants cannot survive in tough environments, but a nest of ants can. What characterizes superorganisms is their intense degree of sociality: their intercommunications, intermetabolisms, and group behaviors. Ants rely on continual exchanges of pheromone signals, they rely on each other to obtain and process food resources, and they do crucial things together, such as engaging in territorial conflicts, which none of them would do individually. Human societies are superorganisms in this carefully defined sense as well. Regarding Gaia as some sort of immense superorganism or super-superorganism cannot work too well, since Gaia's complex systemic interdependency does not rise to the level of genuine sociality as defined above. Avoiding exaggerations of Gaia as any kind of organism would be scientifically prudent. However, it remains the case that biology, while authoritative upon living organisms, may no longer enjoy exclusive say over where living processes may be found. From science's broadest geophysical perspective, Gaia evidently displays a few systemic living processes, although it should not be treated as a unified organic individual.

Skepticism towards the Gaia theory secondly arose because anthropomorphic traits were rashly ascribed to Gaia by some eager environmentalists. It has become common enough to hear environmental enthusiasts—but not scientific environmentalists—say things like "Gaia seeks its own survival," "Gaia protects itself against harm," "Gaia tries to repair damage to itself," and "Gaia makes Earth optimal for life." By assigning humanlike aims and abilities to an entirely natural entity, these claims about Gaia assign teleological purposes to something that cannot really have them. Sufficiently complex systems can display naturally functional features over time, conditions permitting. Yet something functioning in a certain reliable way does not in any sense mean that it is trying to function in that way. Perhaps it is understandable why environmental enthusiasts would find teleological and anthropomorphic accounts of Gaia compelling. As a social species, we are used to dealing with other living agents, and we easily attribute feelings and drives to things like ourselves. Urging harmony with, and protection of, something like a person can engage peoples' instinctive sympathies and helping motivations. However, the scientific reality is the Earth's biosphere cares nothing for itself or anything else, and Gaia certainly could not forestall its doom in any great astronomical or human-caused catastrophe.

The third exaggeration of the Gaia theory inflated Gaia beyond personhood into a mythological entity with conscious, spiritual, and/or divine aspects. Many environmental enthusiasts and some scientists have encouraged this quasi-religious attitude towards Earth. Again, the rhetorical, cultural, and even political advantages to an alliance between environmentalism and religion can be appreciated. On religion's side of such an alliance, prioritizing reverence for the natural world and what is going on in this earthly life couldn't hurt. However, the geophysical and ecological sciences won't permit any personification or spiritualization of what is only a natural system. The living earth evolves to maintain itself as a whole, not to sustain any particular species, no matter how special that species may imagine itself to be. This is a hard lesson for any self-important species to learn, but we must learn it well.

Life has proven to be a resilient feature of this planet. Where there is abundant life, however, there is also death and extinction.

Further Reading

Baxter, Stephen. *Ages in Chaos: James Hutton and the Discovery of Deep Time*. New York: Forge, 2006.

Cattermole, Peter. *Building Planet Earth: Five Billion Years of Earth History*. Cambridge, UK: Cambridge University Press, 2000.

Chambers, John, and Jacqueline Mitton. *From Dust to Life: The Origin and Evolution of Our Solar System*. Princeton, N.J.: Princeton University Press, 2013.

Cohen, Richard. *Chasing the Sun: The Epic Story of the Star that Gives Us Life*. New York: Simon and Schuster, 2011.

Finocchiaro, Maurice A. *Retrying Galileo, 1633–1992*. Berkeley: University of California Press, 2005.

Fortey, Richard. *Earth: An Intimate History*. New York: Random House, 2009.

Kölbl-Ebert, Martina, ed. *Geology and Religion: A History of Harmony and Hostility*. London: Geological Society of London, 2009.

Lovelock, James. *The Ages of Gaia: A Biography of Our Living Earth*. Oxford: Oxford University Press, 2000.

Merchant, Carolyn. *Radical Ecology: The Search for a Livable World*, 2nd edn. London and New York: Routledge, 2012.

McPhee, John. *Basin and Range*. New York: Macmillan, 1982.

Primavesi, Anne. *Sacred Gaia: Holistic Theology and Earth System Science*. London and New York: Routledge, 2013.

Rosen, Steven. *Essential Hinduism*. Westport, Conn.: Greenwood, 2006.

Rosenberg, Gary D. *The Revolution in Geology from the Renaissance to the Enlightenment*. Boulder, Col.: Geological Society of America, 2009.

Spellman, Frank R. *Ecology for Nonecologists*. Lanham, Md.: Rowman and Littlefield, 2008.

Walker, Gabrielle. *Snowball Earth: The Story of a Maverick Scientist and His Theory of the Global Catastrophe That Spawned Life As We Know It*. New York: Random House, 2007.

Worster, Donald. *Nature's Economy: A History of Ecological Ideas*. Cambridge, UK: Cambridge University Press, 1994.

Zalasiewicz, Jan, and Mark Williams. *The Goldilocks Planet: The Four Billion Year Story of Earth's Climate*. Oxford: Oxford University Press, 2012.

3

THE EVOLUTION OF LIFE

What is the most special feature of our home planet? Looking at Earth from an astronomical vantage point deep in space, one might say that the abundant water all over the surface is this planet's most interesting feature. Looking deeper, if a geologist were asked, the answer might be that oxygen is a very peculiar feature. About half of the earth's mass is supplied by oxygen atoms, not just in the water, but mostly in its rocky materials from crust to core, and oxygen's abundance in the atmosphere is also unusual. Plenty of carbon is in evidence, too—and the way that all this carbon gets warmly transferred all around the planet is quite noticeable. But any objective study of our home planet must settle on life as this rocky and wet planet's most interesting feature.

Curiosity about Life's Origins
There are three essential questions to be asked and answered about life on Earth. First, where did all the life that we presently see come from? Second, looking far back in time, how did the first life on Earth come into its existence? Third, why does life take all the diverse forms that so abundantly inhabit nearly everywhere on the planet?

Humanity has long been interested in answering these questions. Stories about the origins of living things and especially of humans are central to most mythologies. Curiosity about life has hardly been inspired exclusively by religion. How life began has been among the most profound questions that religion, philosophy, literature, and science can explore. After thousands of years of imagination, reflection, experiment, and reasoning, our species is still debating and learning about our own origins. Naturally our greatest interest has been ourselves: how and why do human beings have life? Just being able to even ask such questions as an intelligent species gives us a sense of awe and wonder, even while we realize how much more there is to comprehend. Curiosity about this question represents both our capacity for humility and our tendency towards hubris—we see that we are but a small part of the abundant life all around us, yet we judge ourselves to be the most important kind of life on earth. Our kind, our species, seems to be unique in its curiosity about our origins and our distinctiveness.

Curiosity about life is a core trait of religious mythology all over the world. We know very little about the specific views of religions before the invention of writing, but we can make some inferences from the preservation of religious art dating as far back as 25,000 years ago or more, and also from the survival of indigenous societies whose way of life is probably similar to that of much of humanity tens of thousands of years ago. It does appear that humanity all over the world has long been generally convinced that life is just too special and complicated to come from anything less interesting than life. In religious myths, there is always something responsible for making the first living creatures, and this creative force is something that has its own life. The basic idea at the primeval root of all such myths is that life must always come from life.

For example, the extraordinary lives of gods and demigods, some of them very much like people (with superpowers) and some of them more like animals (also with superpowers), are offered by many religious myths all over the world to explain how the rest of life came into existence. The inspiration behind Christianity's creationism is this notion that life had to be specially created by a superpowerful agent. Other religions that don't appeal to divine agents making living creatures still appeal to life's own creative powers, by simply presuming that life has always existed as a fundamental power and simply regenerates itself into new life over and over again. The religious notion of reincarnation is rooted in this view of life as endlessly victorious over death, since death is just a radical transformation of life and not the end of life entirely. By extension, anything closely similar to life, such as a dead corpse of an animal, rotting plants, or even moist soil has been viewed as a potential source of life. In the early times of our species, some observant people must have noticed how new plant life can come from the seeds of dead plants and it can also just grow up out of soil infused with dead plants. It has long been a widespread view across many human societies that matter that has been life or close to life can itself produce life under favorable conditions.

Looking across the long history of religion, two different versions of this root idea that life only comes from life predominate in mythologies, with many religions making an appeal to divine creation, and many other religions appealing instead to regenerating life-matter. It probably is no coincidence that religions of cultures relying heavily on hunting or herding have tended to appeal to divine creation in their mythologies about life, while cultures relying heavily on gathering and agriculture have tended to instead attribute life to regenerating life or near-life. And more recent cultures, such as the proto-European civilizations dating from 1000 BCE and after, which use all of these techniques, typically have complex mythologies displaying both modes of explanation for life.

At the heights of ancient Greece's intellectual superiority, roughly during the period of 450 to 250 BCE, several Greek philosophers pondered the origin of life. Their speculative theories were typically odd mixtures of borrowed religious notions of divine creation and regenerating life-matter, along with some empirical observations from animal husbandry and agriculture and purely intellectual schemes that seem very unscientific from a modern perspective. We may take two of the greatest Greek philosophers as examples of quite opposed theorists about life: Plato and Aristotle.

PLATO

I think, therefore I...
Wait, that comes later...

Plato was an excellent mathematician and geometer for his day. Evidently impressed by the way that mathematical and geometrical proofs conveyed both accuracy and certainty (if done well), he regarded these subjects as finer exemplars of what knowledge should be like. In some of his influential writings, Plato emphasized how mathematics and geometry could be responsible for conveying reliable and unwavering truths. He was especially concerned with the precisely defined things that mathematics and geometry talk about, such as numbers and figures, and with the precisely structured arguments about how to mathematically combine numbers and geometrically construct figures. Plato identified all those precise ideas (or concepts, as we might also say) of numbers and figures as the key feature permitting such precise reasoning about numbers and figures. After all, if numbers or figures were only vaguely conceived, as if different people had differing notions of what the number 10 is, or what a square is, then people would not be able to agree about any inferences or conclusion. Precisely uniform concepts permit rigidly reliable arguments and deductions to truths that everyone can agree with. Plato was not wrong about this crucial role of precisely defined concepts in sound inferences and deductions. Indeed, all logic confirms this.

Plato also leaned heavily towards the view that the best knowledge turns out to be about what is most real. This has been a widely shared view among philosophers and scientists ever since. However, Plato went further down this road of appreciation for knowledge and concepts, to additionally propose that since precise concepts permit the best knowledge of truth, then the best knowledge available to humans must be about those precise concepts. Plato occasionally suggests that the most real things are numbers and figures and other things most like them: all the precisely fixed and rigid concepts we can think of. Plato labeled such concepts as "The Forms." This proposal takes rationalism too far. Plato started from a valid notion that reasoning is about concepts and their relations, and then he went further not only to say that concepts are the best guide to reality, but also to suggest that concepts themselves—the Forms—are the most real things of all. Some enthusiastic followers of Plato, in their establishment of the tradition of "Platonism," basically argued in this manner: reason is about concepts, and reason discovers true reality, so therefore concepts are the true reality. No proper philosopher would exactly argue in such a simplistic fashion for Platonism. In particular, the three-step argument just sketched commits a logical fallacy. Yet the overall trend of

Platonism was obvious. The more precisely defined and permanently rigid something is, the more real it must be.

Plato's philosophical speculations about the Forms and the later Platonist movement might have stayed only a fringe metaphysical perspective on reality. But Platonism was swept up into Christianity during the second and third centuries after Christ, as this new religion's theologians looked for ways to describe the supreme mind of the Christian Creator God. Christians also noticed how Plato identifies a divine "Demiurge," in his book Timaeus, who is credited with intelligently designing and shaping all things in the universe from the chaotic and formless matter that was available at the beginning of the world. A common theme running through Christian theology by the fifth and sixth centuries, and sustained by the medieval Roman Catholic Church, effectively supposed that God created everything in the heavenly and earthly worlds according to perfect forms decided upon by God.

Although the Genesis stories in the Old Testament credit Yahweh for creating plants, animals, and humans, the Bible nowhere states that the species of life could never change or make new species. All the same, dully literal readings of Genesis Chapter 1 and its repetition of the phrase "according to its kind" was forced into the Platonic mold of thinking. For example, Genesis 1:25 reads, "God made the wild animals according to their kinds, the livestock according to their kinds, and all the creatures that move along the ground according to their kinds." Even though this mythological story does nothing more than point out what is obvious to anyone, that the living things we see around us now are members of various kinds of life, Christian theologians added unchangeability to their "essences" without any justification. Lacking any interest in studying life itself, theologians turned to their Bible and Plato for guidance on the long history of life on earth.

Medieval Christian theology continued to stubbornly declare that we could rationally understand how most everything in this world has a God-given and unchanging form responsible for defining whatever it truly is. Each human being, for example, shares in this one true and unchanging form of humanity and each person can only have small particular deviations in appearance from that human form. We are only accidentally our own individuals with our unique features, and those features don't define who we "really" are—we all are essentially one thing, a human being. Every other form of life similarly has its own form defining it as a kind of plant or animal. The biological term "species" has its own origins in this medieval notion that every living thing is a member of a "specific" kind, defined by an unchanging form. We will encounter this medieval notion of biological species again when we come to a discussion of Charles Darwin and evolution. In the meantime, Platonism's obstruction to biology was joined during the medieval era by another obstructing philosophy.

Plato had one great rival during his lifetime: his own student, Aristotle. Aristotle regarded the study of life, not the study of mathematics and geometry, as the most significant field of knowledge. Aristotle had little use for Plato's theory that those Forms are more real than anything on earth. He did agree with Plato that each kind of life had something definite and unchangeable about it, so he perpetuated the notion that each species was fixedly determined by its own distinct and unchanging "essence." Where did this essence come from? Aristotle did not agree with Plato—or many of his fellow Greeks—that humanlike gods were responsible for creating life's many forms. Aristotle did suppose that some supremely powerful and creative force, the "Prime Mover," is causally responsible for structure and general ways of the world. However, this Prime Mover is not directly responsible for the specific things and life forms that populate the earth today.

Instead of crediting a god for making life, Aristotle preferred the rival theory, that life originates in reproduction from life (with the transmission of tiny living eggs or seeds) and also from warm, wet matter close to life such as carcasses along with soils, shores, and bottoms of ponds where dead organisms decay. Aristotle thought that the "vital heat" or "living soul" still retained by those earthy materials is responsible for the spontaneous generation of some lower kinds of life such as insects. This label of "spontaneous generation" applied by later Aristotelian commentators does not intend to imply that life arises by pure chance or arises in an instant. Here, "spontaneous" only means that no living organism is responsible for the emergence of life in such a hypothetical case, and that the near-life earthly matter can accomplish this "birth" entirely on its own.

Where Did Life Come From?

This theory of spontaneous generation lingered alongside Christianity's theology in later European culture. The very old notion of life's earthly reincarnation and regeneration had survived in common mythologies alongside Christianity. Some intellectuals additionally had access to Aristotle's writings, which Europeans rediscovered in the twelfth and thirteenth centuries, long after Islamic scholars had thoroughly digested Aristotle. Christian theology then accommodated and absorbed many of Aristotle's philosophical speculations while retaining its mythology of a supremely intelligent God who created the first kinds of life. As a consequence, the notion that every living thing is a member of a species that has never changed since creation only became more entrenched.

During the period when modern science was emerging in the 1500s, these two religious ideas of divine creation and fertile matter were commonly accepted all over Europe. When medicine revived an interest in life's reproductive abilities and novel theorizing got underway in the 1600s, one theoretical controversy dominated all others: Does life only originate in birth from living organisms, or does some life get spontaneously generated from near-life earthy matter? Advocates of living birth ridiculed the theory that things like insects can be born directly from a carcass or from wet soil without any other living organism involved. Advocates of spontaneous generation appealed to the "authority" of Aristotle, still favored by the Church, and also to such "observations" as seeing tiny insects crawling out of rotten meat or dirt. During a time before the invention of the microscope, how could this question about the origin of life be answered? Could life come from nonlife?

Italian naturalist Francesco Redi (1626–1697) performed simple experiments which demonstrated that rotten meat would not generate any maggots, or anything else alive, if flies were prevented from landing on the meat. However, he showed how flies permitted to land on the meat were responsible for laying the eggs that produce maggots, which in turn produce flies. The publication of his experimental results, and further accumulating evidence from similar experiments after his own, gradually eroded away the popular notion that insects, snakes, or frogs could originate in nonliving matter. Louis Pasteur (1822–1895) later demonstrated that not even bacteria could originate from nonliving matter. In his experiments with flasks of liquid broth favorable to bacterial life, he showed that a sealed flask of liquid that was first boiled to kill any bacteria would not have any bacteria in it later on, no matter how much time passes. But soon after a boiled flask had its seal broken, bacteria were found growing in the flask's broth. Evidently bacteria could travel short distances through a broken seal and eventually reproduce inside the flask's liquid.

By the eighteenth century, biology was able to break free from many of the theological chains preventing sound experimental science. Even the Christian notion that humans are utterly different from all other animals was crumbling. After all, early scientists curious about life had plenty of life besides humans to observe and classify. The naturalists who explored the earth and collected all kinds of plant and animal creatures for further study were among the earliest biologists. Carl Linnaeus (1707–1778) is regarded as a pioneering biologist who accelerated botany (the study of plants) and zoology (the study of animals) with his careful classifications of species. His flexible version of a taxonomy that names a species with its genera and species (for example, Homo sapiens) became the standard taxonomy, along with his use of the broader categories of kingdoms, classes, and orders. Linnaeus's accurate judgment that Homo sapiens should be grouped together with the other primates such as chimpanzees and gorillas provoked a theological controversy. He resisted protests demanding that humans should be classified quite separately with a superior status. Linnaeus could find no reason to suppose that human beings are demeaned or diminished by his zoological classification, and he stated his own view that humans have good company since animals have souls too.

Linnaeus is an example of a biologist who retained some Christian ideas about the existence of the immortal soul, even while the science of biology continually advanced to situate human beings alongside the other animals. By the time of Darwin in the mid-nineteenth century, biologists agreed that every species of life resembles some other species, since they can be grouped together into larger common classes, orders, and genera, and biologists also agreed that every living thing is descended by reproduction from an earlier living thing. Looking backward into historical time, some biologists like Darwin were ready to discard the Bible entirely in order to guess at the natural origins of the first kinds of life from which all the rest of life descended. Biology had been founded on answering the question "could life come from nonlife?" in the negative. However, two centuries later, biologists were starting to raise a somewhat different question: "Did the first life come from nonlife?" Answering this question without relying on a Biblical God now required an affirmative answer. Even Darwin was forced to ponder this ultimate question of the earthly origin of all life.

The right answer by contemporary biology to the simple question, "Does life come from nonlife?" combines both traditional notions about life born from life and life generated from nonlife. All current life on Earth comes into existence by getting "born" in reproduction (either asexually or sexually) from other organisms. No life comes from nonlife today. However, when the earth was still young, the very first forms of life arose from complex chemicals and compounds, not themselves alive, which gradually acquired the ability to sustain their own existence, organize into replicating forms, and survive and grow over time. In a way, the very first life forms did spontaneously generate from materials that were not alive. The process of life's earliest origins could not have been "spontaneous" in any purely chancy or instantaneous sense—the first sorts of primitive life did not simply "jump" together with no cause at all. Biology's account of life's origins describes a long history of many interlocking chemical processes that all obey the basic laws of thermodynamics and chemistry. Current incomplete portraits of life's complicated origins seem anything but "spontaneous," so we will drop that outdated term. Instead, biology talks about "abiogenesis"—the origin of life from nonliving organic compounds.

Biology cannot avoid the task of eventually explaining in detail how simple life could arise from nonliving chemical compounds. This is so, even if earth's life did not actually start on the earth. The intriguing idea that some sort of extraterrestrial life fell to earth long ago, finding a hospitable environment and jump-starting life on earth, still assumes that this life had an origin somewhere else. It does appear that many kinds of organic compounds were distributed around the early solar system and would have fallen to earth in meteorites. It is presently unknown whether such organic compounds would have supplied key materials for life not already available on the early earth, or whether fully living organisms could survive interplanetary space or the fiery descent to earth. It is possible, perhaps, that very primitive forms of life first emerged on our planetary neighbor Mars, and got carried to Earth when rocks got blasted off Mars by an asteroid collision. Only future discoveries about the possibility of early life on Mars could help support that hypothesis.

All the same, somewhere and somewhen, life started from nonlife. The Big Bang theory of the universe does not make any room for the possibility that something alive was among the energies erupting at the universe's start. If life did not entirely start here, it did have to start in abiogenesis somewhere that provided ample chemistry and heat. Because we know very little about the ancient environments of other planets, at least compared to the modest knowledge we now possess about the earth's own history, biology might as well try to figure out how earthly life could have arisen here. This effort would not be a waste of time. Even if future evidence points towards a nonterrestrial source of the first life in our solar system, a workable theory of life's origins on a planet like the young earth would still be useful. Therefore, we will continue to talk about biology's attempts to account for a terrestrial origin of life.

Biologists have received crucial assistance from other scientific fields, especially geology and chemistry. Late nineteenth century geology supplied more accurate estimates of the earth's age in terms of billions of years rather than thousands of years. Darwin's theory of natural evolution required very long stretches of time for new species to evolve, so an extremely old earth provides the sort of deep time needed for evolution. Geology also became confident that the earth began as a very hot ball of molten rock, and the crust required several hundreds of millions of years to cool so that surface temperatures could drop below the boiling point. There was a long duration of time when there was no life on earth. But there would have been plenty of CO_2 and organic compounds, along with hot water and bright sunlight. Geologists also came to understand how the earth's crust is continually in flux with drifting continents and active volcanoes, helping to account for the constant replenishment of carbon on the surface and in the atmosphere.

Chemists determined that organic compounds are all built from carbon atoms, and explained why carbon is essential to life. Carbon easily bonds with more carbon and other common atoms available on the earth's crust, and especially with those useful for life: hydrogen, oxygen, nitrogen, sulfur, phosphorus, magnesium, and potassium.

There is no other atom quite like carbon. Carbon can tightly bond with almost any other element, it can supply the skeletal chains for very long and complex molecular compounds, and its versatility permits over ten million different organic compounds. Life also relies on the unusual properties of water for good reason: water loosely bonds with the widest variety of other atoms and molecules and also functions as a solvent capable of dissolving other molecular bonds. The liquid into which organic compounds can most easily disperse, stay dissolved, and mix together at a wide range of temperatures is water. Geology and chemistry proved critical for providing background information about the conditions under which life could start and survive.

Basic organic compounds will easily form under a variety of environmental conditions, including the conditions present on the early earth's surface. The early earth's atmosphere lacked free oxygen, but there was plenty of nitrogen (N_2) and carbon dioxide (CO_2), and water (H_2O) was also abundant. Free oxygen is very damaging to organic compounds because it is so reactive; life protects itself from oxygen as well as selectively uses it for respiration. By the way, this explains why life would emerge but once on earth. We couldn't see life getting started now because any novel organic chemistry wouldn't survive the oxygenated atmosphere, even if it wasn't promptly digested by bacteria. Additional simple organic compounds also littered the early earth's surface, having been brought down by meteorites. Many kinds of organic compounds were chemically interacting according to the laws of thermodynamics on the early earth. Researchers are pursuing a variety of theories that the building blocks of life—the hydrocarbons, lipids, amino acids, proteins, and eventual nucleic acids that can sustain continuous metabolism and reproduction—could form near available energy sources.

Charles Darwin suggested a hypothesis about the energetic conditions for life's start that remains one of the leading options for biology today. He wrote to Joseph Dalton Hooker in 1871 to suggest that life began "in some warm little pond, with all sorts of ammonia and phosphoric salts, light, heat, electricity, etc., present, that a protein compound was chemically formed ready to undergo still more complex changes." Darwin couldn't have known that ammonia was actually an unlikely presence on the early earth, nor could have Stanley Miller and Harold Urey, whose 1952 experiment with a warm mixture of simple molecules produced amino acids such as glycine and alanine. Recent experiments with a more accurate mixture of simple compounds also produced many amino acids, including simpler nucleotide versions of nucleic acids, the essential enzymes controlling organic chemistry. The beginning of abiogenesis would have been the chemical formation of the first self-replicating organic molecules, such as simple ribonucleotides, that could take advantage of surrounding energies to last for significant amounts of time, at least longer than a few seconds.

One currently prominent theory identifies longer ribonucleic acids (RNA) as the original and primary replicators, with the more complex deoxyribonucleic acid (DNA) taking charge of preserving genetic information later on in evolution. Life's origins could center on the development of RNA, since even short strands of RNA would have had the capacity to help synthesize both protein structures (for the metabolism of surrounding energy) and orchestrate the formation of more RNA (for the reproduction of itself). Today, some kinds of viruses only use RNA. During abiogenesis, the most effective RNA molecules would generate, with mutations and accidental synthetic reactions, ever more complex RNA over time. Complex RNA would contain the first genes consistently directing even more complex structures of proteins and lipids to perpetuate and protect metabolism.

An RNA's "metabolism" during abiogenesis was the utilization of surrounding energies—heat, light, other organic chemistry, oxidizing mineral compounds, etc.—to sustain a process that, among other byproducts, results in a copy or near-copy of that RNA. It is very hard to say what kinds of "metabolisms" would have been possible then, but much interest is focused on adenosine triphosphate (ATP). ATP is at the core of the Krebs cycle, discovered in the 1930s by Hans Krebs. All life uses the Krebs cycle to chemically react with available molecular compounds (the "food") to tear them apart to obtain the immediate release of energy, promptly used for building other organic compounds necessary for maintaining life. Organic chemists are trying to theorize how ATP would be built up from its components. It is already clear that ATP's components would not be hard to produce where there is plenty of nitrogen, water, and CO_2 along with the sorts of phosphates, based on phosphorus, that come along with lava breaking through the crust. With something like ATP working to break down some energy sources and produce more organic compounds in the process, more complex organic compounds such as RNA can then be constructed. Indeed, the earliest RNA compounds would have proven useful for harnessing ATP work, keeping it focused on available energy sources. The Krebs cycle by itself just catalyzes everything in the vicinity, including neighboring Krebs cycles, and quickly sputters out, reduced to its simpler compounds once again. A Krebs cycle that accidentally produced some proto-RNA which in turn helped the Krebs cycle continue longer would be a very special kind of organic chemistry, signaling that abiogenesis was well under way.

Once RNA was taking advantage of the Krebs cycle to both keep the Krebs cycle going and replenish itself, a small nest of recycling organic reactions could sustain the processing of available energy sources over time. It is still too soon to declare that life has now arisen, in the fullest sense of life—this stage is still in the middle of abiogenesis. Matters are still too chaotic. No genuine species would exist at this early stage of abiogenesis, since any RNA copying would be imperfect. Any RNA's "code" would fluctuate too much over time, so there wouldn't be a large number of RNA "colonies" having much the same stable RNA. It is also too soon to be speaking of evolution during abiogenesis. However, there was plenty of raw competition for survival. Better success at using enough energy to be constantly replicating is a dynamic and creative process yielding many diverse "offspring" over time.

One major problem for any primitive life to solve is raised by the way that ordinary organic chemistry will unhelpfully tear apart anything approaching life. The special organic chemistry that can consistently perpetuate its own reproduction needs protection. Life got its proper start when replicating organic compounds used their own membranes to protect a regular inner metabolism from the wilder external chemistry, and the first cells were established. Cell membranes, at first just single layers of lipids, permitted protection of the Krebs cycle and the master RNA/DNA code from distortion and consumption by neighboring RNA processes. Before cells, the top threat to an RNA's survival was the RNA copy it just produced, which would treat its own "parent" as just convenient organic chemistry to be consumed. After cells were constructed, neighboring cells wouldn't prioritize eating each other, and cellular life soon preferred colonies of species, just as we observe today.

Rival theories for the origin of life agree on the elemental conditions favoring abiogenesis, but these theories offer different locations or energy sources for fueling and protecting life. Fueling life requires energy gradients: places that are far from thermodynamic equilibrium because there is a constant source of fresh energy, in the form of raw heat or chemical reactivity. Ocean shores and tidal flats could have supplied clay molecules for concentrating complex amino acids and proteins. Another theory gaining respect locates the origin of life near deep sea underwater vents and volcanoes. Where lava meets ocean, alkaline phosphates, sulfides, iron compounds, and other minerals abruptly meet CO_2-infused acidic seawater. The mineral deposits can build up in irregular ways to make tiny cell-sized chambers where organic chemistry could accelerate. These sorts of deposits can be found in rare places even today. In those chambers, there is plenty of hot water and also alkaline-acid energy gradients, where intense chemical reactions are quickly rearranging free protons and electrons between compounds in a never-ending pursuit of lower energy levels. All that hot chemistry can supply both the organic compounds needed to build ATP and the Krebs cycle essential to abiogenesis, and also the excess energy required to keep building them.

Could abiogenesis and life have begun deep underwater? Fossils of bacteria 3 billion years old or more have been discovered near these kinds of intensely hot lava vents. Such strange environments are still teeming with life, as many varieties of small organisms thrive in the presence of noxious gases and sulfur dioxide. The ability of single-celled life to survive in such extreme environments of temperature, pressure, and chemistry cannot be doubted, as these "extremophiles" can be still found today living in places that would quickly kill any other life.

If abiogenesis did begin very early in the ocean's depths, safe from heavy asteroid bombardment and intense surface volcanism that ended 3.8 billion years ago, then the road to life is almost as old as the earth itself. Fossil evidence for early abiogenesis is unavailable, since soupy ferments of replicating organic compounds would not leave substantial traces lasting billions of years. By the time that life figured out how to utilize membranes, and mineral deposits were shaped by cellular life, some sort of fossilization would be possible. The earliest undisputed fossils are of bacteria from around 3 billion years ago, and there are signs of life's traces in rocks going back to 3.5 billion years ago or more.

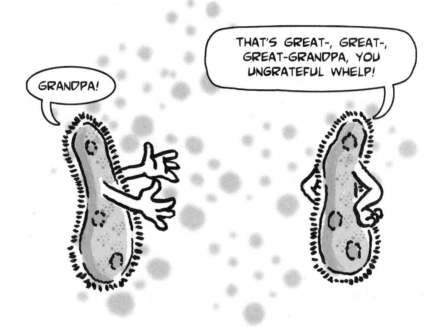

These periods of time are set apart by geologists. The earliest is the Hadean eon, from the earth's origins around 4.5 billion years ago to 3.8 billion years ago. The Hadean eon was followed by the Archaean eon, from 3.8 billion years ago to 2.5 billion years ago, and then came the Proterozoic eon from 2.5 billion years ago to 542 million years ago. Coordinating geology with biology, it is estimated that the long period of abiogenesis was concluded by 3.5 billion years ago. Abiogenesis had come to its fruition when the first true single-celled organisms were using stable RNA and DNA to sustain internal metabolisms for transforming energy and replicating as species. That period of abiogenesis also produced abundant loose RNA and DNA that kept replicating by invading living cells and parasitically using those cells' own machinery for replication. These parasitic replicators lacking their own metabolism are called viruses. Viruses have survived magnificently, going wherever cellular life went to cover the planet. During the rest of the Archaean eon and much of the Proterazoic eon, life consisted entirely of viruses and single-celled organisms, although they were becoming somewhat more complex during that period. Only towards the middle of the Proterozoic eon, about one billion years ago, did multicellular life proliferate.

Simple single-celled organisms, along with viruses, could have been the only forms of life that ever emerged on earth. But more complex life having vast diversity did eventually evolve from these simplest kinds of life, to take advantage of many different kinds of environments and energy sources across the earth. How did all that complexity and diversity come from such primitive origins?

How Did Life Evolve?
Charles Darwin rightly judged that all life on earth is descended from one original species of simple cellular life. This hypothesis is called the "common ancestor" hypothesis. Darwin's judgment was well-informed by three empirical facts known to biologists of his time. Two more crucial facts about life's hidden genetics were discovered in the twentieth century to additionally support the common ancestor hypothesis. First, biologists long understood how every species of life resembles some other species, since each species can be grouped together with similar "cousins" into larger common genera, orders, and classes. Second, nineteenth-century biologists could see through microscopes how all life on earth uses the same basic plan of cells that have similar internal structures. Third, biologists and paleontologists could look at the way that older fossils typically have simpler body structures than younger fossils, showing how early life was quite simple, yet capable of having progressively more complex descendants.

Just these three pieces of evidence alone suggested to several naturalists before Darwin, including Georges-Louis Buffon (1707–1788) and Darwin's grandfather, Erasmus Darwin (1731–1802), that earth's life had descended from one original life form whose descendants gave rise to a variety of organisms modified to survive in different kinds of environments. During the twentieth century, biologists additionally discovered how all life uses the same basic mode of ATP chemistry to extract energy, and also how all life uses RNA and DNA in almost exactly the same way to produce amino acids and proteins and to ensure orderly self-replication.

These five facts are core reasons why descent from a single common ancestor is by far the best explanation for the kinds of life that have existed on earth. The odds that two different original life forms contributed to life's genetic diversity are calculated to less than one in a million. Common sense also suggests that the first form of cellular life would have had a huge advantage over any life form later emerging from abiogenesis. Any emerging candidate for the second life form would have to compete for resources with the far more efficient first form of life, and it would soon be eaten by the first anyway. Furthermore, biological explanations for the way that life has evolved shows how the numerous variations among organisms that we see today could have gradually come from one single original life form.

Evolution is not the same as any mechanism responsible for causing evolution. Determining that all life on earth descended with changes from a single common ancestor is one thing. Knowing how a species can gradually change to produce another species is quite another thing. Biology offers naturalistic explanations for the process of evolution, and those explanations are empirically confirmed by detailed evidence from the study of fossils and living species, and the expanding knowledge of genetics. Darwin is rightly credited for scientifically establishing his theory of evolution by the mechanism of natural selection. Darwin was not the discoverer of evolution itself, as already noted, but his theory of natural selection was the first good theory to explain how evolution can happen over time to produce the variety of organisms seen in the fossil record and observed on earth today. Darwin's theory of evolution by natural selection identified the primary cause of evolution of life since its origins, and it was his theory that made the modern science of biology possible. Since Darwin, biologists have discovered additional mechanisms, such as genetic drift, that also are responsible for some kinds of evolutionary changes. Those are additional natural causes for evolution, and they are not any sort of "refutation" of Darwin's theory of natural selection. They are only discoveries that natural selection does not act alone to produce evolutionary change. Only natural causes are ever responsible for evolutionary changes to life.

Because the continual copying of DNA and RNA in every cell is an intricate series of chemical processes, disruptions to that genetic code or "mistakes" in copying are always possible. Such mutations can be caused by natural background radiation, invasive DNA such as viruses, and the like. Even in the earliest species of single-celled life, minor genetic variations would be present across the species so that very few individual cells would have identical DNA. There are other causes of genetic variation across a species, such as genetic drift, gene flow, and epigenetic mechanisms. Over stretches of time, these kinds of variations in genes can cause significant changes to a species's heritable traits and hence be responsible for evolution. However, the ultimate long-run imposition of species change was usually natural selection.

The idea behind natural selection is relatively simple: since environments supply finite resources and dangerous hazards, and there are variations throughout a species, some members of a species will get more resources than others. Enough differences in resources generally results in different reproduction rates. Unless organisms are surviving well enough to pass along their specific traits, those traits will be less represented in future generations. Those having advantageous traits will have enough resources for creating offspring, so they will pass on more of their traits to the next generation than those which don't have such advantageous traits.

This mode of evolution of natural selection proposes that nature itself effectively "selects" which organisms do well enough to survive and pass on their genes and traits. Over long periods of time, a species tends to increase its fitness for success in an environment, so long as that environment remains suitably stable for that species. If the environment is changing, a species must adapt or perish. Most species have perished. Over 99% of all the species that have ever lived on the planet have gone extinct. They were forced to evolve into a different species or (far more often) they weren't able to adapt fast enough to survive.

Some species do not change much while their required environment reliably supplies enough resources. Many species of bacteria and other microorganisms, for example, are not significantly different from their ancestors hundreds of millions ago. They have few requirements, needing only some water and organic matter, available from their environment to survive well. There are a few examples of plant and animal species that have changed little from their ancient ancestors, only because their required environments have similarly been reliably supportive. However, most of the earth's surface has generally been a constant scene of change, often dramatic change, and natural selection is therefore responsible for much faster evolution. It should also be emphasized that life itself has been responsible for some of the most dramatic changes to the earth's environment.

The long history of life on earth displays how biological complexity is very difficult to achieve. However, when a novel form of life appears, it can cause immense changes to the earth. From the time of life's origin at least 3.5 billion years ago, life remained very simple for the next 1.5 billion years. Life consisted only of simple single-celled organisms (setting aside the questionable viruses), which are classified into the archea and bacteria domains. Single-celled life had plenty of environmental challenges to survival, apparently leaving few resources for much innovation. The early earth continued to be dominated by extreme conditions, where boiling hot waters were full of toxic salts and acids, and high-energy ultraviolet light glared down. Many of the archea capable of tolerating such extreme conditions left descendants down to this day, and can be found thriving where no other form of life possibly could. Most microorganisms are bacteria. There is hardly a location on the planet lacking millions of bacteria per square inch. Bacteria have been one the greatest ecological forces on the planet ever since, perhaps only rivaled by volcanic activity. Without the impressive evolution of bacteria long ago, the earth would have become a dead planet like Mars instead of the lush vibrant planet we enjoy today.

Perhaps life began deep underwater, but it did reach the shallows of ocean shores and lakes, relying on sunlight and carbon dioxide to power its metabolism. In these half-lit warm waters, impressive innovations caused huge jumps in complexity. The first dramatic innovation arrived around 2.7 billion years ago, when some bacteria started using the sun's light to reverse the Krebs cycle so that it combines water and CO_2 to assemble organic molecules called carbohydrates. This process of photosynthesis by cyanobacteria uses kinds of chlorophyll, aided by energetic photons, to break apart water molecules and then recombine water's hydrogen with CO_2 to produce carbohydrate chains plus some free oxygen as a by-product. Photosynthesis produced hundreds of times more energy than anything life had been doing so far. Cyanobacteria promptly flourished in warm waters around the planet, soon overwhelming the other kinds of bacteria.

Free oxygen in the atmosphere is highly reactive and deadly to single-celled organisms lacking the protections that cyanobacteria had to erect. We can see the corrosive effects of free oxygen whenever we leave a piece of fruit exposed to air, soon turning it brown, or when we see how fast most metals rust when exposed to the weather. The non-photosynthesizing bacteria either died off, hid deep underground away from any air, or adapted to the oxygen. Some bacteria that did adapt began living alongside cyanobacteria in layered cooperative colonies called microbial mats. Their fossilized remains are among the kinds of stromatolites that can still be observed today. Other adapting bacteria eventually tried to "eat" cyanobacteria by engulfing them. Stealing carbohydrates is easier than making them, but cyanobacteria could not have been easy to digest. Within another billion years, some cyanobacteria species developed an unusual symbiotic relationship with other bacteria, permitting themselves to live entirely within their hosts. This kind of relationship, called "endosymbiosis," has happened several times during the long course of evolution. Today, a plant's cell contains smaller cells called chloroplasts responsible for photosynthesis. Animal cells also contain smaller cells, called mitochondria, which are also descendants from independent bacteria that found life more congenial inside their engulfing hosts.

Symbiosis and Complexity

Symbiosis proved to be the greatest innovation for life on earth. Microbial mats were the first major symbiotic development, dating back to at least 3.5 billion years ago. Three additional kinds of major symbiotic relationships evolved during the period from 2 billion years ago to 1.4 billion years ago: endosymbiotic and eukaryotic cells, sexual reproduction, and multicellular life.

Early complex cells, called eukaryotes, evolved as early as 2 billion years ago, and were flourishing around the planet by 1.5 billion years ago. These eukaryotes also began storing their primary DNA inside a new inner membrane to form the cell's nucleus. Around that same period, from 2 to 1.5 billion years ago, a second kind of cooperative symbiosis, sexual reproduction, was also starting in many species. Explanations for the evolution of sexual selection are not well-established. Sexual reproduction involves blending the DNA of two organisms to produce a third organism with its own unique DNA. Why would an organism benefit from having a uniquely unpredictable DNA? We have already mentioned some ways that different kinds of bacteria were symbiotically surviving together for mutual benefit. However, any species might benefit even more by taking more than its fair share of resources, becoming a parasite upon its neighbors. With cellular life becoming more interdependent and symbiotic, species needed defenses to obstruct parasites, and viruses remained a permanent problem as well. Unique DNA would be a major advantage for confusing potential parasites and hostile viruses. Additionally, sexual reproduction increases mutations and hence variation within a species. There are also obvious disadvantages too. Most mutations are just deadly or quite disabling. Payoffs for sexual species must be large. Under certain conditions, such as intense competition with neighboring species or pressures to adapt to changing environments, sexual reproduction proved to be advantageous in the long run for many eukaryote species, which were flourishing by around 1.2 billion years ago.

The next kind of symbiotic cooperation was the development of genuine multicellular organisms. Evidence for multicellular organisms goes back to around 2 billion years ago. These larger complex organisms were flourishing by 1.2 billion years ago, and they mostly used sexual reproduction. Although multicellular forms evolved independently many times in a variety of species from cyanobacteria, molds, algae, and sponges to fungi, plants, and finally animals, the multicellular design remained a major problem for evolution to solve each time. The lengthy epoch of time required for multicellular organisms to abundantly evolve indicates that it is not easy to obtain cooperation from all of an organism's cells to reproduce offspring.

The basic problem of multicellularity is that the cells of a multicellular organism have to be usefully differentiated into distinct kinds and functions, implying that only some of the organism's cells supply the genes that are passed on to offspring. These special germ-line cells are responsible for the traits of the next generation, and the rest of the organism's cells face only death. Why should some cells cooperate, only to fail evolution's basic test of passing on their genes? Still, from the perspective of the genes inside those cells, those genes are getting indirectly copied, so there is a kind of reproductive survival for them. If cells A, B, and C all have the same genes M and N, but only cell C is used to produce another organism, A and B eventually die but at least the genes M and N in cells A and B have been reproduced, indirectly, when cell C passes along its copies of M and N to the new offspring organism. In effect, if the work of all the cells can be coordinated, then the work contributed by cells A and B to the organism is rewarded when offspring have their genes. From the perspective of genes, multicellularity usually makes sense if it results in better survival and more offspring. However, making sure that cells A, B and C stay cooperative over the organism's lifetime was a vast evolutionary hurdle, and many species solved that problem in a variety of ways too numerous to mention here.

Keeping matters in perspective, only a tiny fraction of species made the leap to multicellular form, as most of the rest of life remained viral and bacterial. However, those multicellular species became the most interesting life forms on the planet. The photosynthesizing plants grew larger to capture more sunlight while the hungry feeders became better consumers, and they all used their bigger size to prevent being eaten themselves. Many multicellular feeders figured out how to move around, becoming the first animals around 800 to 600 million years ago. By 600 million years ago, the world's oceans and lakes were full of microbial mats, algae, fungi, and sponges, along with primitive versions of worms, slugs, arthropods, and strange kinds of small animals difficult to classify. With advanced cellular differentiation, some cells became devoted solely to coordinating sensory cells with motor cells, and the first nervous systems were activated. Animals sampled the waters they were immersed in, sensed the ambient organic chemicals around them, began navigating their environments, and became better at pursuing prey and avoiding dangers. Intelligence was born.

The "Cambrian explosion" of novel kinds of aquatic animals from around 550 to 530 million years ago shows how evolution by natural selection can produce rapid diversity. During the period of the Cambrian explosion, the basic body designs of all major groups of animals were established, including that of humanity's phylum Chordata, which includes all animals such as vertebrates having a characteristic nerve cord running the length of the body. There was nothing really so abrupt or spontaneous about this "explosion," however. The same evolutionary forces and laws were at work, as they always had been since life's origin. Nothing unnatural was involved at all. We must remember how fast evolution can happen when environments dramatically change for species. Furthermore, this "explosion" could have started earlier than 550 million years ago, as very soft-tissued small animals were developing into the bigger complex species seen in the Cambrian fossil record. It is difficult enough to find fossils of larger and tougher animals from so long ago, and nearly impossible to find any traces of very small and soft animals dating before 550 million years ago. Such animals were little more than squishy bags of water, membranes, and proteins (even present-day jelly fish have tougher structures), and they would be promptly consumed so that nothing was left behind to fossilize after their death. Only a few rare imprints of traces and trails of these tiny soft animals have survived for examination, but there is good reason to suppose that those early animals were evolving in interesting ways too.

During the Cambrian explosion, the earth was undergoing major climate changes, causing some advantageous conditions for all life, such as increased oxygen levels. Furthermore, the way that animals had been moving about their surroundings was itself a modification to the environment. The environment for a tough-skinned animal that can burrow down into the seabed, or can swim above the seabed, or can perceive its prey with eyes, or can clutch at prey with jaws, or can protect itself with calcified armor, is navigating a very different environment from what surrounds an animal that has a soft body, a slow crawl, only dim sensation, and weak suction. Two quite different animals can physically be situated in the same place, yet inhabit two very different habitats. An environment is not the same sort of thing as a habitat— the organism, in a way, determines its habitat. Smart animals like birds, beavers, and humans deliberately reshape and construct some of their habitats.

During the Cambrian explosion, surplus energy and natural selection favored animals trying radically new ways to gain advantages over their food, and over each other. Not only were their physical environments opening up, but their competition was changing as well. As soon as one species developed workable eyes, many more species developed eyes and other defenses too, since organisms had to adapt to more dangerous prey that was now eyeing them. No sooner than biting jaws developed, other species developed armor of spines, shells, or scales. Biologists refer to these "move and countermove" developments as an evolutionary "arms race." The Cambrian explosion stopped when the various kinds of new animals had exploited all the novel aquatic niches available, and the surviving species had achieved a higher-level standoff in the predator-prey balance.

By 530 million years ago, large animals were crawling and swimming around with facility. Underwater plants underwent some fast evolution during this time as well, developing vascular structures to become bigger and tougher. The only remaining sort of environment to be exploited by life was dry land. The timing was excellent. Only cyanobacteria and a few kinds of microbial mats had crept up onto shorelines before then. Once again, free oxygen in the atmosphere made all the difference. Oxygen had been building up in the atmosphere for two billion years, and higher concentrations of oxygen permit a stable ozone layer of O_3 molecules in the atmosphere. Ozone quickly reacts with metals to make rust, just as O_2 does, but it can last at the highest levels of the atmosphere, where its unusual properties make it ideal for just one thing: capturing any high-energy radiation that hits it from outer space. A thick ozone layer protects earth's surface from almost all of the life-damaging radiation from the sun and other cosmic sources. By 480 million years ago, lichens and tiny fibrous plants were gripping the rocky shorelines, and arthropods and worms followed that food chain.

Hey Frank, you need to come see this…

Plants gradually colonized the earth's land, developing stalks for height, roots for nutrition and stability, fibrous defenses against cold temperatures and water loss, and many kinds of leaves for capturing more sunlight. The long Carboniferous period, starting 360 million years ago, was the age of vast forests. Animals kept following the plant life—this truly was the age of insects. But they had company. Some species of shallow-water fish with adaptable fins for crawling and primitive lungs for breathing air were spending time on dry land by 390 million years ago. By 300 million years ago, the age of the amphibians and reptiles was under way. Only the evolution of warm-blooded mammals from early reptiles around 200 million years ago, and birds from later dinosaurs around 150 million years ago, remained to complete the list of major kinds of animals capable of exploiting the rest of the environmental niches on the earth's surface.

Only vast humility can do justice to the long history of evolution that eventually produced large omnivores such as Homo sapiens. Our genes carry much of that long heritage, which we share with much of the rest of life: genes that keep the Krebs cycle going; genes that build our cellular membranes and tissues; genes that permit us to digest all kinds of foods; and genes that build our muscles and nervous systems. Humanity shares 30% of its genes with yeast fungus, 70% of its genes with the sea sponge, 82% with the platypus, and 98% with the chimpanzee. Life on Earth is truly one large family.

Misconceptions about Evolution

Immediately after Darwin brought the theory of evolution by natural selection to the attention of the scientific community, it was challenged by both scientists and by nonscientists. Scientific advances in geology, paleontology, and genetics have only strengthened the firm conclusion that life has evolved since its simple origin, and that natural selection has been the primary long-term factor responsible for the evolution of species from earlier species. Theory has become established fact. The species observed today evolved from earlier species, and natural selection has helped shape the ways that evolution has occurred. Biologists continue to learn about the specific details of ways that species have evolved, and the ways that species are genetically related to other species. Scientific disputes necessarily arise over the best interpretation of evidence for rival explanations for specific events in the course of evolution. But this is all a matter of normal scientific research entirely grounded in the solidity of evolution by natural selection.

Any rejection of evolution or natural selection arises from a poor understanding of science, or from some entirely unscientific notions. Resistance to the acceptance of evolution of inherited traits by natural selection has three sources: (1) simplistic cognitive biases against a theory like evolution; (2) inadequate ideas about how science works; and (3) religions preferring a creationist explanation for life. Dealing with cognitive biases first is the best way to then refute religious challenges to evolution. Explaining science has to take human psychology and enculturization into account.

The way that science describes how it works doesn't automatically fit with basic cognitive frameworks. Science also has to deal with the unintended consequences of its own public relations campaigns. For example, science has long prided itself on justifying theories with observable evidence, and justifying a particular theory on the grounds that it is the simplest theory which explains the most evidence. People unfamiliar with scientific method apply their own "commonsense" notions to think about these two criteria. For example, it is easy to ask skeptical questions such as "Is evolution even observable?" and "Why can't we see evolution happening today?" If these questions aren't satisfactorily handled in ways that ordinary people can understand, their third question is often something like, "Why couldn't God have made all creatures exactly the way they are?" An appeal to God can seem to many religious people like the simplest yet most comprehensive answer to life. How can biology supply a more reasonable alternative to this sort of nonscientific yet commonsense reasoning? Each sort of question deserves a patient answer. Let's start with simple cognitive biases against evolution.

"Is evolution even observable?" It is not unreasonable to expect observable evidence. Like asking whether the earth's rotation is visible or whether continental drift is happening, an affirmative answer about evolution depends on proper perspective. Fossils deeper in geological strata are both older and simpler than younger fossils and today's life as well. There are few exceptions, only where recent but simpler organisms are explainable by degenerate evolution under unusual habitat conditions. If someone denies that fossils are very old—implying that geology knows nothing—then observing old fossils seems impossible, but this is a case of deliberately blinding oneself, and can't be science's fault. Similarly, tracking the genetic similarities among living species to find out which cousins descended from an ancient ancestor is a way to indirectly observe evolution. Someone believing that genetics knows nothing will be blinded to evolution again, but this is not science's fault either. For those able to perceive and understand the geological and genetic evidence, evolution is quite easy to see. Creationists aren't merely challenging a single theory about evolution—they have to dismiss several fields of science together, explaining why creationism appeals to people refusing to accept anything science says.

"Why don't we see evolution today?" Actually, evolution can be observed nowadays if you know where to look. The careful study of any population of organisms in a species reveals the expected genetic diversity among individuals, showing that the variation in genetic inheritance is abundant. Species are changing their genetic codes and bodily traits in the wild, right in front of us. It is also easy to produce new hybrids of plants simply by growing them near each other. The way that humans have deliberately bred novel varieties of dogs and grains just by regulating how "nature takes it course" also shows how species will diversify and change. Although the longer stretches of time typically required for species evolution make it difficult to directly observe, the evolution of one species from another parent species has been noticed. Just a few examples can be mentioned here. Species of European wildflowers called Tragopogon or goatsbeards were brought to America in the early 1900s, and after hybridizing with local wildflowers, a new species related to the goatsbeard but unable to cross-fertilize with the original species had evolved after a few decades. The fruit fly species *Rhagoletis pomonella*, which originally infested hawthorn trees, divided into two species during the nineteenth century so that the second species could infest apple trees instead. Other fruit fly species, and the much-studied E. coli bacteria, have been carefully observed to divide into separate species under laboratory conditions as well.

"Why do species today stay the same?" The plants and animals we can easily see don't change much because each species has "earned" its place of reliable survival within its own ecological niche to which it has gradually adapted. Under changing or challenging environing conditions, significant evolution within a species can happen rapidly—noticeable changes can happen within hundreds of years, if not faster, in specific cases. The development of a new species by natural selection usually requires more time, typically on the scale of many thousands of years at minimum. However, cases of very rapid evolution of new bacterial and micro-organic life have been observed (such as drug-resistant bacteria). The same is true even for a few larger plant and animal species as well. Back when Darwin had to defend his theory of evolution almost by himself, the religious authorities relied on Plato, Aristotle, and the Church to deny that species could change. Fortunately, odd convictions that had nothing to do with any actual natural evidence were not permitted to stand up to science for long.

"Evolution means abrupt big changes to bodily form, but there's no evidence for that." This objection comes from a failure to appreciate how much time is involved with evolution. The fossil records for many varieties of plants and animals do display dramatic changes in bodily form, changes that sometimes take place within a few thousand years. That's pretty abrupt from the perspective of geological time over hundreds of millions of years. If this question is instead demanding that evolution shows its evidence for one species suddenly changing into a very different species within less time than that, the answer is that neither the fossil record nor genetics offers that kind of evidence. Fortunately, the theory of evolution by natural selection does not assume that species evolution at that fast pace has to happen, so biologists don't need such evidence. There has been plenty of deep geological time to permit regular slow evolution of gradually changing traits, occasionally punctuated by faster evolution (such as the Cambrian explosion) to produce the life forms seen in the fossil record and to produce the paced transitions in life's genetic codes.

"Each species looks so carefully designed to survive, so it's smarter to credit God." The way that a species operates to "fit" its own required environment can look designed. Each species is a marvel of complexity in its own way, from the bacterial parasite to the bald eagle. However, that "design" is actually the result of a very long process of evolution in which organisms accidentally found a way to survive better in an environment and pass on more of their traits than less-successful members of its species. Repeat this natural selection upon the small variations in a species over thousands of generations, and large variations build up to produce new species after new species, each one making a better "fit" with what its environment has to offer. Great complexity has arisen in species this way. Even the traditional example of the eye, truly a complex organ, has been found to have gradually evolved in the fossil record by many stages from very simple membrane folds accidentally more sensitive to light. Furthermore, evolution predicts what creationism cannot, that a species will retain features that do not perfectly suit it to its current environment. And that prediction is fulfilled everywhere. Every species retains minor features and useless genes that have nothing to do with its current environment, and they all suffer from some features only marginally managing to function well enough to let the species survive. Evolution does not produce perfection, but only good enough features to produce the next generation. That is why all life dies—surely that is a major design "flaw." Humans are not exempt. We have all sorts of vestigial "design" flaws, from the useless appendix in our guts to our awkward knees in our legs. Just the poor design of human knees, easily damaged from frequent use or awkward movements, would hardly be the top choice of a designer making humans fit for upright walking or running. Evolution predicts imperfection, and that prediction is abundantly confirmed.

"How could mere chance throw together something as complex as an organism?" When the ordinary mind tries to imagine chance producing something complicated quickly, it naturally can't. Our minds assume that only chance comes from chance. Fortunately, evolution does not require believing that mere chance is responsible for life. Mere chance is one thing, while useful accident is another. Strictly speaking, life is a kind of chemistry, and nothing in chemistry happens by pure chance. Strict laws of physics and thermodynamics dictate what can happen under set conditions and what cannot. The ways that the first organic molecules can work together to harness heat and light to assemble even more complex organic compounds have nothing to do with chance, and everything to do with natural necessity. The only question is how much mixing and time is available; and there was plenty of that. By useful accident, certain steps were added to that basic chemistry at some point during all that mixing, which proved irreversible: once those additional chemical reactions got started, they would continue so long as the basic environing resources remained plentiful. A useful accident is a novel way of doing's life's advanced chemistry that makes this way of living possible longer into the future. All the same, a happy accident is not mere chance—a happy accident always obeys natural laws—but it is accidental in the sense that life did not strictly have to evolve that way. Genetic mutations that further modify life's traits are the outcome of ordinary physical and chemical laws, but a few accidentally improve fitness for survival. Evolution by natural selection is not perfectly predictable, since the vast scale and complexity of dynamic organism-environment transactions transcends any mathematical calculation, but that is far from relying on pure chance. The laws of physics and chemistry describe what can happen thermodynamically within a given short period of time, but they don't exactly dictate what must happen biologically over vast stretches of time.

Just a few ordinary cognitive biases, so prevalent in human psychology, can prevent people from grasping why evolution happens. Evolution can even explain why we have those cognitive biases. For getting along in daily life, we can rely on what is immediately observable, we can expect natural causes to be obvious and impressive, we need only take seriously what happens quickly while we can watch it, and we explain complexity using our inborn design-detectors, so useful for figuring out each other. Troubles involving cognitive biases arise when we apply intuitive common sense to unfamiliar natural events that are far vaster and older than the human scale and time frame.

Even if these cognitive biases are managed and set aside so that scientific explanations can be understood, a modest amount of scientific knowledge can still set up huge barriers to following evolution. Here are a few examples:

"If life began with simply chemistry, why hasn't it been synthesized in the laboratory?" Theories for the stages of abiogenesis are still quite tentative. All the same, most of the chemical steps from simple organic compounds up to amino acids and replicating proteins have already been sketched out. Many of those steps have been duplicated under laboratory conditions. However, doing all of them, doing them in the correct order, and performing that entire synthesis under the right controlled conditions, will require decades more research. It is unreasonable to expect a laboratory to do in a few years what abiogenesis managed to produce during hundreds of millions of years. Proving that we understand the precise chemical conditions on the early earth will require vast effort. Still, it is hardly foolish to predict that self-replicating and metabolizing forms of life will be artificially synthesized before the twenty-first century is over.

"Why does life violate the second law of thermodynamics?" Actually, life never violates any laws of thermodynamics—the essential chemistry of life is the fulfillment of thermodynamics. From a narrow perspective, life seems to grow more complex and energetic, but from the widest thermodynamic perspective, life is just a phase of a wider ecological system of heat exchange among the earth's core heat, the sun's hot radiation, and the cold depths of space. The earth as a whole is shedding excess heat into space, in obedience to the second law's dictum that entropy must increase by transferring energy from a warmer to a cooler place. In thermodynamics, one must take all energies into account, so that only the entire earth-sun-space system is the closed system. The earth's surface is an open system, and open systems can display small local concentrations of complex energies without violating thermodynamics. In fact, for a complex open system such as the earth's surface with all its excess heat, thermodynamics predicts dynamic and partially chaotic energy exchanges that efficiently shed additional heat to keep increasing entropy for the earth-space system as a whole. And life is precisely one of those kinds of energy exchanges. From a thermodynamic perspective, life's energies are just swirls of wet combustion supplying the earth with an additional way for energy to flow up and away from the crust. On a planet like the earth, life is practically necessary, in blind obedience to the laws of thermodynamics.

Where's Link?

"Why aren't there more transitional fossils?" This question exposes ignorance about fossils. To have as many fossils as we do requires geological luck, since plant and animal bodies are easily destroyed by other life, erosion from weather, erosion in the soil, and geological forces. Furthermore, in a sense all fossils are transitional fossils, just as all species today are transitional species. No permanent species has ever lived—in the short run, every individual organism is a bit of a mutant, and in the long run, all species are either transitioning out of existence from extinction or transitioning on to a new species. The notion of a transitional fossil is not a scientific notion, but a metaphor used to popularize paleontology while scientific knowledge of the history of life gradually grows. In other words, we seek the "transitional fossils" between dinosaurs and birds, not because the dinosaur species or the bird species were somehow the "true" or "fixed" species that were supposed to live and something else had to bridge them for a while, but just because we already have fossils of some species, and we would love to see more fossils to link them in a genealogy.

"Why are there still so many missing links?" Like the idea of a "transitional fossil," the idea of a "missing link" is not a scientific term. It is another popular phrase describing how paleontology seeks more fossils to fill in developmental gaps between the many fossils we already have. Evolution is already obvious from the fossils now filling the world's museums. To complain that evolution can't be right because there aren't enough transitional fossils to fill all the missing links is a complaint made from ignorance about how paleontology works. In any case, paleontology does find more and more missing links all the time. When creationists are never satisfied with more missing links showing up to bridge the gaps between humans and hominid ancestors like Homo erectus, and between all hominids and ancestral cousins to the chimpanzees before that, it becomes obvious that no amount of discovered links will satisfy these self-blinded deniers. Only a stubborn commitment to denying that humans are descended from anything not human lies behind this ignorant skepticism.

Life Is Naturally Precious
Understanding how evolution has shaped life makes it more special than creationism ever could. According to creationism, a supernatural God designed each and every species, and created them on the earth, ready to live in their assigned habitats. Creationism does not deserve to be labeled as a theory, or even as an explanation, for creationism explains nothing.

Perhaps an incompetent and uncaring Creator, or a pantheon of imperfect gods, would create such imperfect and constantly adapting species, but contemporary religions ignore those options. If God is so great, as the world's monotheisms declare, then the imperfections of evolution and life are left unexplained. And God's intentions must be questioned, for designing so many horribly infectious, parasitic, and lethal organisms that torture the rest of life. If God is using evolution to slowly create intended creatures instead of just flashing the final products into existence (an odd enough notion in itself), then the incredible amount of suffering, destruction, and death littering evolution's meandering path is all blood on God's hands. Is God just a monster? Some theological-minded creationists offer artificial answers to such questions, and the debates over "the problem of evil" ensue. Since those debates have nothing to do with the validity of scientific evolution, they can be passed by.

Other creationists don't worry about evil, instead wanting God to have some role, any role, in evolution. So desperately do they seek something for God to do, that they imagine God occasionally tweaking evolution with genetic mutations to ensure that the "right" species—such as humans, of course—eventually evolve. Since there is zero evidence for divine intervention in either the fossil or genetic record, and no supernatural explanation is needed in any case, this sort of interventionist creationism again fails to offer an interesting explanation, much less a testable theory. Only mysterious claims about unverifiable miracles are really being made by creationists. That is why these "evolutionary creationists," such as the Intelligent Design defenders, are usually reduced to complaining that biology has yet to completely explain some feature or another of organisms. This "God of the gaps" strategy—when science can't explain X yet, insert "God did it"—is not anything like a respectable intellectual effort. The medieval theologian William Occam was enough of a philosopher to realize how an unnecessary hypothesis that can't really explain anything should be cut away and dropped.

There is no appropriate role for a Creator in scientific evolution at all. Only religious believers, already convinced that there must be something so special about us that only a God would make us, would

Occam ponders... as do a lot of us

be driven to blend science and religion. Any synthetic blend or mutual accommodation can only betray and violate the science of evolution.

Natural evolution does not regard any species or any individual as special at all. Evolution did not "intend" to produce us, or any other species. The only thing that all living species necessarily have in common is that they have managed to survive so far. In a sense, they all are "winners" in the struggle for survival. And they all, sooner or later, will be losers. All life must be sacrificed, sooner or later, to more life—and suffering is essential and important to this process. All life must struggle, die, and get eaten. Furthermore, no species is really the "purpose" or "point" of evolution: evolution is a grand thermodynamic process without aim or direction.

There really isn't anything in particular about life that evolution is progressing towards. Take intelligence, for example. Animal intelligence is special, in a sense, because it is rare among all the species that have ever lived on earth. But the emergence of animals came about through a series of useful accidents that did not have to happen. High intelligence, so prized by us humans, is nothing that special either. Having big energy-consuming brains is just one strategy among a great many for the survival of a species. And it is not clear that it will pay off. Very few species have ever gone far down the path of relying on high intelligence, among the many millions of species that have ever lived. If high intelligence was so great for survival, the dinosaurs would have evolved it among some of their species—they had 200 million years to do it. Mammals did evolve high intelligence, at least a few species of mammals did, after some 150 million years on the planet. Mammals didn't do it under the stress of surviving the tyranny of the dinosaurs. And mammals only did it a few times since their liberation, in dolphins, primates, elephants, and so forth. We ourselves have risen to such heights of intelligence that we can clearly see that nothing had to turn out this way. But it did, and we may be grateful for the happy accident.

Evolution is the simplest, dumbest thermodynamic process capable of creating ever more wonderful and endlessly beautiful forms. Evolution is spectacularly wasteful and extravagantly accidental. In the long run, evolution only maximizes entropy, which is the long road to equilibrium and death. Hopefully that is a very long road. Evolution is the only thing that can make a planet shine more interestingly than any sun.

Further Reading

Benton, Michael J. *The History of Life: A Very Short Introduction.* Oxford: Oxford University Press, 2008.

Browne, Janet. *Darwin's Origin of Species: A Biography.* London: Atlantic Books, 2013.

Coyne, Jerry A. *Why Evolution Is True*. New York: Penguin, 2009.

Dowd, Michael. *Thank God for Evolution: How the Marriage of Science and Religion Will Transform Your Life and Our World*. New York: Penguin, 2008.

Fry, Iris. *The Emergence of Life on Earth: A Historical and Scientific Overview*. New Brunswick, N.J.: Rutgers University Press, 2000.

Gargaud, Muriel, Purificación López-Garcìa, and Hervé Martin, ed. *Origins and Evolution of Life: An Astrobiological Perspective*. Cambridge, UK: Cambridge University Press, 2011.

Kitcher, Philip. *Living with Darwin: Evolution, Design, and the Future of Faith*. Oxford: Oxford University Press, 2007.

Margulis, Lynn, and Michael Dolan. *Early Life: Evolution on the Precambrian Earth*, 2nd edn. Sudbury, Mass.: Jones & Bartlett Learning, 2002.

Marcus, Bernard. *Evolution That Anyone Can Understand*. Berlin and New York: Springer, 2011.

Parker, Andrew. *In the Blink of an Eye: How Vision Sparked the Big Bang of Evolution*. New York: Basic Books, 2009.

Scott, Eugenie. *Evolution vs. Creationism: An Introduction*. Berkeley: University of California Press, 2005.

Shanahan, Timothy. *The Evolution of Darwinism: Selection, Adaptation and Progress in Evolutionary Biology*. Cambridge, UK: Cambridge University Press, 2004.

Sherratt, Thomas N., and David M. Wilkinson. *Big Questions in Ecology and Evolution*. Oxford: Oxford University Press, 2009.

Wilkins, John S. *Species: A History of the Idea*. Berkeley: University of California Press, 2009.

Yarus, Michael. *Life from an RNA World: The Ancestor Within*. Cambridge, Mass.: Harvard University Press, 2010.

4

HUMANITY'S BEGINNINGS

All around the world, religions tell tales about the First People and how they were created. Many religions add details to their human creation stories to try to explain differences among people. Religious myths recount tales about why there are males and females, what formed people into separate tribes and races, and how the first people acquired different languages. These myths are often innocent tales that imaginatively entertained people who had no better way to think about things long ago. Some aren't innocent at all. Myths from many religions, including the world's largest religions, also perpetuate cultural prejudices and social injustices down to this day.

The history of our evolutionary origins from the primates tells the true story of how humanity began, and what the first people were like. Additional scientific accounts trace the cultural differences of groups of humans as they migrated into nearly every region of the world over the past 70,000 years. Science explains how we are all Africans by descent, and why our superficial differences are so small compared to our similarities. The scientific understanding of humanity's beginnings and what humanity is basically like can disprove mythological notions and throw doubt upon ugly prejudices, bigotries, hatreds, and other kinds of religious excuses for oppressing people.

The First People

What might have the first human beings been like? They left behind a few fossilized remains of their skeletons and skulls, and more are discovered almost yearly. We can tell that they are human like us, because their overall body shape, skull size, and facial features are just like ours, and not like other primates or the found evolutionary links between humans and primates. In terms of biology, their anatomy makes them one of us, Homo sapiens. But what we are most curious about is this: What were they like? If we could somehow watch them, what would they be doing?

They were probably a lot like today's San peoples of southern Africa. The San are one of the few remaining indigenous hunter-gatherer groups on the planet. Their simple foraging way of life closely resembles that of the first Homo sapiens. These early humans evolved into a distinct species around 200,000 years ago in sub-Saharan regions of eastern and southern Africa. Genetic studies show how the San are closely related to their neighbors, the Khoe. However, the Khoe-San as a group are much less genetically related to the rest of humanity today. Their genetic heritage diverged from the rest of the human gene pool around 100,000 years ago. Actually, it would be fairer to say that the rest of humanity diverged away from that early gene pool.

There were few humans 100,000 years ago, perhaps a couple hundred thousand in all, but those humans had many genetic differences among them. There wouldn't have been enough differences to be very noticeable, but they were sufficient to give Homo sapiens a fair amount of diversity, despite the way that almost all were still living in sub-Saharan Africa. Such diversity would be expected, since Homo sapiens had evolved from an earlier Homo species just 100,000 years earlier, and still carried many genes from that older heritage, as well as newer genes distinguishing Homo sapiens as more intelligent and adaptable. The Khoe-San have more genetic diversity among themselves than the diversity found among other human populations around the world. They have kept that greater diversity because their ancestors remained in sub-Saharan Africa, and it was that group of sapiens which retained the most diversity in their gene pool.

To represent the idea of a gene pool, geneticists sometime use the idea of a large jar of marbles having dozens of different colors. These colors represent genes, the special genes that made us distinctly human around 100,000 years ago. You and I don't have all those genes today. That's because we descended from small groups that migrated away from sub-Saharan Africa long ago. Let's suppose you are descended from a group that left Africa, and their descendants kept heading towards southeast Asia. Your earliest ancestral group that diverged from sub-Saharan sapiens took some of that genetic diversity with them, but only some. Imagine randomly pulling out 20 marbles from that large jar of marbles, to stand for your ancestral group. These 20 marbles won't display all of the colors that the marbles inside the jar display. By analogy, the amount of genetic diversity represented by your marbles is less than the original amount of genetic diversity that all Homo sapiens had 100,000 years ago. The lesson is that whenever a small group of sapiens migrate away from a larger group, they take only a portion of that larger genetic diversity with them.

The Khoe-San have the most genetic diversity seen in any group of humans alive today. That is because they are more directly descended from the original sapiens that remained in sub-Saharan Africa. Later human divergences away from the Homo sapiens gene pool carried less genetic diversity with them than what stayed behind. This makes sense, because small human groups traveling away from the regions of humanity's beginnings only take a portion of that gene pool with them.

Having traveled a much shorter distance from humanity's beginnings, in both a migratory and genetic sense, the Khoe-San haven't moved far in a cultural or linguistic sense, either. The Khoe later added pastoral practices where herding and grazing was possible, but many of the San never changed their basic way of life. They primarily live off the land through hunting and gathering, and they nomadically travel across the countryside in extended families called bands. There are fewer than 100,000 San left today. Some, like the Khoe, have been absorbed into the work of farming. The languages of the San and the Khoe have been affected by this gradual absorption. The Khoe-San do still use the "click" languages of their region. Besides the consonant

I think I've lost my marbles...

sounds we use, like the b or t sounds, they add tongue clicks, such as our "tsk, tsk" sound for disapproval. These languages appear to be extremely old and complex, but clicking consonant sounds were not carried beyond sub-Saharan Africa by migration. Later languages intermingled with the Khoe-San languages, so it may prove impossible to uncover what the earliest human languages were like.

The Khoe prefer for themselves the name Khoekhoi, which means "the real people," indicating a very narrow perspective. This name also conveys a sense of superiority over the San, whose name in that same language means "the outsiders." What the San originally preferred to call themselves has been lost to history through the loss of their own linguistic traditions. The label of "Bushmen" by white colonizers is just as crude. Similarly, the second-oldest branch of humans, the "Pygmies" of the rainforests of central Africa, mostly use Bantu-related languages, and even where older languages survive, original meanings of tribal names are uncertain.

Where the meanings of tribal names for aboriginal groups are preserved in their own languages, those names frequently mean "the people," "the first people," "the people of the land," or the people of some prominent geographical feature like a lake or a mountain. The San tell a tale that the first people came up from a large hole in the ground, a hole supposedly in their ancestral lands, though none of the San can find it today. Other indigenous peoples in southern Africa, such as the Tsonga and the Xhosa, share that same tale about their first ancestors coming up from the earth, while the Zulu and related tribes in that region say that the first humans burst out of growing reeds. Norse legend says that the supreme god Odin created the first man and women from trees native to Scandinavia. Aboriginal tribes of central Australia, among the most isolated on the planet, also regard the first ancestors as fundamentally connected to the ancestral land. In that land, the Warramunga and neighboring tribes tell tales about the first people who lived underground at first, and how their emergence onto the surface long ago brought about creative deeds responsible for rituals and customs.

Similarly isolated from other civilizations until three or four centuries ago, many native tribes of North and South America sustained the hunter-gatherer way of life inherited from their Stone Age ancestors who crossed the Bering Strait by land or by sea at least 15,000 years ago in several migratory waves. The Lakota of the Great Plains have a central myth for the creation of the first people by one of the supreme gods, who places them underground to live until further mythic events bring them up onto the land. The Navajo of the Southwest similarly locate the first humans underground until their expulsion by gods towards their current homeland. The Cubeo of the northwest Amazon jungle attribute the origin of the first people to the emergence of the original tribes from their respective holes in the ground. The indigenous Mapuche of southern Chile, whose name means "people of the land," regard the first people as the earthly offspring of spirits generated in the course of the world's contentious creation out of a first great spirit.

Typical ancient tribal myths similarly describe how the first people in the world happen to be the first people of the tribe telling the tale. It is also crucial for these early myths to explain how these first people were created together with their homelands, fusing their origins and destiny. Across Africa and Asia, up into the Arctic and down to Australia, and throughout North and South America, the few surviving aboriginal tribes regard themselves as the first inhabitants of those lands. Indeed, the term "aboriginal" indicates the first inhabitants of an area, akin to the term "indigenous." The politics of applying those terms put aside, the migratory habits of hunter-gatherer groups makes it difficult if not impossible to confirm many claims to original inhabitancy. Nevertheless, it is typical for these groups to view themselves as not only the first native inhabitants of their homeland, but also the first humans on the entire earth. Their legends commonly relate how their own group was created first, and how other peoples must have descended from theirs. Many indigenous groups may not be wrong about the way they first inhabited their lands, but they are all wrong about being the first humans on the planet.

The way that the oldest cultures typically regard their own peoples as the first humans, so that their ancestors were part of original creation, is entirely to be expected. Dating from a time when small bands of Stone Age foragers were sparsely dotted across the landscape, their stories focus on the explanations for the important natural features around them, and justifications for their own social customs and rituals. Other bands of humans, only rarely encountered, didn't seem to be something requiring special explanation, and the notion that the creation of the world involved the generation of many kinds of peoples didn't occur to many of them. Only later, among early agricultural and pastoral peoples, do narratives about the first humans typically acknowledge diversity and divisiveness among different societies. Not surprisingly, these narratives still credit one's own tribe with the closest ancestral relationship to those first people.

This typical self-centered view of humanity's origins is repeated and amplified by the earliest civilizations, dating from 6,000 years ago. Each of these civilization's mythologies, from Egypt and Sumeria to India and China, recount how the gods created the first people, who also are the first people of that same civilization. Common Indian and Chinese myths older than Hinduism or Daoism recounting the creation of the first humans point to one of the gods as their creator, who shaped people from available earthly things like the dirt, plant life, or other animals. Among the oldest Chinese stories is the tale of Nuwa who created many kinds of animals and then the first humans from clay. The ancient Egyptians preserved several core myths involving humanity arising from the deeds of one god or another. One early myth involving an important deity named Khnum says that he made humans from clay on his potter's wheel. Greek mythology relates how the heroic Prometheus made the first humans from clay.

Egyptian myths become less literal and more abstract, as a priestly class put their own ideas into the tales. One myth dating from the Old Kingdom period (before 2200 BCE) and lasting for millennia say that the first people came from the eyes of a primeval god, such as the tears from the weeping eye of Ra, or directly from the seeing eyes of Amun-Re. The "Instructions to Merikare," supposedly composed by the king of Egypt's capital city around 2100 BCE, recount the official myths of creation. According to this mythology, humans are the supreme god's "images" that came from this god. Having the divine likeness is always a mark of supreme worth. During his reign around 1330 BCE, King Tut's full name was Tutankhamun, "the living image of Amun." The worship of Amun as the supreme Egyptian god was promoted wherever New Kingdom power and influence extended, especially westward to Libya, southward through the Sudan, eastward across the Levant from ancient Palestine to Syria, and northward into Greece.

Among the tribes well versed in Egyptian religion were the nearby Hebrews. In the first chapter of Genesis, the first humans are created "in the image" of the gods (the Elohim). Genesis also relates a second story of humanity's creation in the second chapter, as Adam and Eve, are formed by Yahweh from dust—Adam first, then Eve from a rib of Adam (echoing a Sumerian tale). They are placed in a "garden," a lush area full of fruits and grains, and their descendants were supposed to live off this fertile land. This tale has nothing to do with the early Hebrew way of life as nomadic pastoral peoples living in arid uplands before they settled in the land of Canaan. But it does mimic the Sumerian myth of how the gods created the first humans by molding some clay and then settling them to tend the green fields between the waters. Two of the four rivers flowing through the Garden of Eden according to Genesis are the Tigris and the Euphrates, the two large rivers where the Sumerian civilization arose.

The First Homo Sapiens
There never was a time when there were one or two first human beings. The earliest Homo sapiens numbered in the many thousands, as the result of a gradual evolutionary divergence from a larger and much older species called Homo erectus. Homo sapiens had emerged by around 200,000 years ago in eastern Africa, and the oldest skeletons are found in Ethiopia. There wouldn't have been many of these original humans, perhaps totaling a few tens of thousands of individuals. The way that dramatic climate change had made much of Africa drier and less hospitable probably accelerated our evolution towards even higher intelligence, group cooperation, and better technologies, but it also meant that sapiens remained small in number for a long time. The way that sapiens, like Homo erectus, depended so much on hunting game animals meant that many hundreds of square miles were required for feeding just a small group. Before the domesticating and herding of animals, sapiens had to perpetually track and chase their usually faster prey.

Some groups of sapiens traveled north, reaching southeast Asia (present-day Arabia and Israel) by 125,000 years ago, but they apparently died there or returned to Africa. Some went west into the Congo rainforests, the ancestors of the Pygmies. Many drifted south from Kenya down to South Africa by 115,000 years ago, the ancestors of tribes such as the Khoe-San. Another wave of sapiens journeyed north out of southeastern Africa around 70,000 years ago, and all people not of direct African descent are descended from this migration. Because the non-African portion of humanity has a smaller amount of genetic variety compared to the African portion, the number of initial individuals responsible for this migration may have been fewer than 1,000 mating pairs of parents. By 60,000 years ago they had spread across Asia. They were the first Homo variety to inhabit Australia by 50,000 years ago, and the first into the Americas by 17,000 years ago.

Homo sapiens wasn't the first upright great ape to explore continents. That honor goes to Homo erectus, by far the most evolutionary successful Homo species. If Homo sapiens manages to survive for more than a millions years, sustain itself on this planet in ecological harmony, and give rise to further extraordinary species, perhaps then our species can be judged to have surpassed erectus. From the time of its own evolutionary origins around two million years ago from smaller hominins, it proved to be an extraordinary great ape. Around eight million years ago, there were no gorillas, chimpanzees, or protohuman apes—Homo sapiens could not have directly descended from gorillas or chimpanzees—but only a shared primate ancestor. The ancestors of gorillas divided away first, and then the chimpanzees divided from the hominin line, the line that led to erectus and sapiens, around six million years ago.

The earliest hominids weren't much different from chimpanzees. They were about the same size, had skeletons like chimpanzees, and lived in forested regions. But much of Africa was changing, as the forests were shrinking due to less rain and other effects of climate change. The hominids who spent more time in the drier grasslands next to the forests were slowly becoming different from the primates that remained in the forests. By five million years ago, they were clearly changing.

Their skeletons were a little taller and better for walking on two feet, their feet were less like hands and more like our feet, their skulls were a little bigger for bigger brains, and their teeth were smaller due to a somewhat different diet. Evolution continued to shape successive species towards a more erect height and longer stride. Despite living in more difficult terrain away from the safety of the trees, they were able to survive. Homo erectus was the later beneficiary of that evolutionary trend when it developed around two million years ago. Unlike the other apes, such as the chimpanzee and gorilla, and its own ancestor Australopithicus, Homo erectus was able to survive in several kinds of terrains and climates across Africa, from green forested lands to dry savannahs and rocky coastlines.

Homo erectus had the largest brains, more than 850 cubic centimeters, twice that of gorillas. They had longer legs for walking upright and long-distance running, along with shorter arms with flexible hands for carrying tools. Their tools included thrusting spears and sharp stone choppers and cleavers, and they had the ability to use fire. Cooking foods increased nutrition, saved energy from digestion, and freed calories for larger brains. Because of their ability to travel across wide distances and survive in diverse terrains, some of them migrated out of Africa at different times, depending on favorable climates. Homo erectus was the first species to fully combine a reliance on coordinated hunting of large animals with flexible gathering of many kinds of plant foods. Childhood was prolonged several years, gender roles were intensified, and they were the first to live in the larger groups called bands of up to 50 individuals.

The earliest Homo erectus that stayed in Africa are also called Homo ergaster, but destiny lay with erectus, as its brain size continued to increase up to 1100cc. Erectus populated much of Africa and migrated throughout Southern Europe and Asia, reaching south into Indonesia. Because this species lasted so long, until as recently as 12,000 years ago in China, it had many varieties, and some of these varieties could interbreed. An erectus variety called Homo heidelbergensis emerged around 600,000 years ago in eastern Africa. This variety had grown anatomically closer towards Homo sapiens, both in terms of skeletal stature and proportion, and it also reached our brain size, around 1,350cc.

Heidelbergensis lasted until 300,000 years ago, about the time when further development towards sapiens would have been beginning, but insufficient fossil evidence prevents establishing more than a few details. Some groups of heidelbergensis also journeyed north out of Africa into southern Europe and Asia in another wave of migration during the late Pleistocene age. In Central Asia and Europe, heidelbergensis produced Homo neanderthalensis, and the Neanderthals flourished from around 300,000 years ago down to their extinction in southern Spain around 25,000 years ago. If heidelbergensis did produce both the Neanderthals in Europe and Homo sapiens in Africa, that may explain how some interbreeding was able to happen after sapiens encountered neanderthalensis in Europe around 43,000 years ago, as evidenced by a small amount of genetic commonality traceable in human populations today. Denisova hominins, probably another offshoot of heidelbergensis in eastern Asia, shared some genes with sapiens. Sapiens may have encountered other late varieties of homo inhabiting east Asia. Homo floresiensis, a "little people" clustered in Indonesia until around 10,000 years ago, would have been an offshoot from erectus. Too few remains have been discovered to date, however, and the possibility that skeletal remains are actually from abnormally small Homo sapiens has caused doubt. A significant percentage of shared genes among all these descendants from Homo erectus indicates not merely a shared genetic heritage from the past, but also a capacity for interbreeding as well. There are additional genetic indications that one or more additional Homo species had left genes in Homo sapiens as well. Since species differentiation usually requires two million years of evolution, the many kin of erectus and heidelbergensis were more like varieties of a single species than fully separated species.

Evolutionary destiny, it turned out, lay with sapiens alone, as the rest gradually died out. Sapiens didn't forget them entirely. Tall tales survived to describe curious encounters with near-humans—some large and dangerous and others small and furtive, but all of them primitive in technology and inarticulate in speech. Impressive legends require embellishment long after events that inspired them, as any good story-teller would agree. There are historical and cultural reasons why a fable about a giant is different from a superstition about ghosts, and both are quite different from myths about gods.

The sapiens who reached Europe, after a long journey westward from the rich steppe lands of Central Asia, found equally rich territories for the hunter-gathering way of life. Until a revived Ice Age, that is. As the glaciers descended from the north 30,000 years ago, much of the northern hemisphere down to the Alps and the Himalayas became tundra until around 18,000 years ago, so sapiens had to concentrate in southern Europe and southern Asia. Sapiens may have contributed to the extinction of the last Neanderthals, already diminished by climate as well, by hunting so much of their prey, and possibly clashing directly with them. When the ice had receded northward 12,000 years ago, sapiens expanded again, dramatically repopulating northern regions around the world. Once again, the late Pleistocene hunting and gathering lifestyle resumed from France to Korea, from Scandinavia to South Africa, and more migration into North and South America became possible as well.

Now expert at hunting all kinds of big and small game, as well as fishing and foraging, humanity ate very well during the warm period following the end of the Ice Age. All humans on the planet, numbering probably no more than four or five million by 10,000 years ago, had basically the same intelligence, language ability, physical capacities, and roughly the same skill sets, put to more or less the same uses. And they had the same curiosity about the world around them.

Whether inhabiting India or South America, the Congo or Japan, humans were erecting simple shelters, clothing themselves, adorning themselves with jewelry, creating paintings, sculpting figurines and pottery, and using nets, hooks, needles, blades, axes, and various projectile weapons like spears, slingshots, and arrows. They were living in bands or small clans numbering from 50 individuals up to a couple of hundred people in more favorable areas. Some differences in ways of life had to be sustained, of course, since surviving in quite different

environments required different cultural techniques. But the late Stone Age was an era when everyone on the planet was basically using the same kind of culture. Their lives seem so different from ours. Yet babies born today are born ready to live that kind of life as well our own.

We naturally find ourselves fascinated by our minor physical differences and our obvious cultural differences. Homo sapiens is precisely the sort of intensely social species with large brains capable of noticing and tracking such differences among ourselves. Yet these abilities, our distinctive abilities to scrutinize each other's looks and behaviors, our facility with designing symbols for social distinctions, and our capacity for endlessly talking about each other's social activities, reputations, and rankings, all depend on having the same brains. Those brains have produced remarkable technologies and social structures since those Stone Age days. Yet these advances are due to greater population sizes and the accumulation of cultural learning. Our advances during the past 20,000 years aren't due to any major biological modifications to how we socialize, communicate, or sustain culture.

The Myth of Adam and Eve

Homo sapiens as a whole is astonishingly similar from a biological point of view. The small amount of variability among populations decreases as the walking distance from Africa increases. Our differences are still quite interesting from our own perspective, of course. Notable divergences include the ability to easily digest some foods such as milk, resistance or susceptibility to some diseases, minor facial features, and skin colorings. These small genetic differences mostly have to do with group adaptations to regional environing conditions during humanity's migrations around the globe. Some mutations make it easier to digest foods only available in certain regions, or to resist a disease found in some part of the world. These are the sort of minor variations within a species that evolutionary genetics would expect. These genetic variations are so minor that it usually is not possible to blame them for health problems or shorter lifetimes. One's actual environment, diet, and lifestyle today, and not any environment or way of life thousands of years ago, plays the overwhelming causal role in enjoying good health or suffering from poor health.

It is also dangerous to overstate the degree of genetic variability between continents, since there is typically far more variability within a population on one continent than there is between them and the residents of a different continent. Furthermore, the genetic variability presently found within a population, such as some small groups living in sub-Saharan Africa, can be much greater than the genetic variability measured across a large population. There are a small number of distinctive genetic differences that can be traced back to their mutation origins in certain geographic territories. Yet that divergent ancestry is the only thing involved, not any innate divisions within the entire gene pool of humanity. There are no races of humanity, and the only remaining meaning to "race" is political, because that term isn't a scientific categorization and it fails to represent anything natural to our species.

The notion of human races is demolished by genetics. The way that sapiens as a whole is almost genetically identical has to do with the way that all human populations arose from a small number of early sapiens in eastern Africa around 100,000 years ago or less. This is a very short period of time from an evolutionary perspective. Mutations do happen to any species' DNA as long periods of time pass, but these periods are extremely long. Estimates vary, but at least two million years are usually needed for major variations in a species to diverge enough to form two distinct species. Sapiens has been evolving across the planet for just 5% of that vast amount of time, so there hasn't been enough time to even evolve significant minor variations. Biologically, there are no subspecies or any other kind of innate subdivisions to Homo sapiens.

We aren't clones of each other either, of course. The small number of mutations within the species are solid information for science. A few minor differences to DNA in the nucleus and the mitochondria of a human cell can tell a story about the travels of one's ancestors. If a certain mutation happened in a population living in the Middle East 50,000 years ago, for example, that mutation would be carried by descendants of that population wherever they went, so people in Spain, Finland, Siberia, and South America would also carry that mutation, since migrations out of the Middle East populated those other areas in stages. If a population in Central Asia 40,000 years ago had another certain mutation, both mutations would be found in any populations descended from that region, such as Spain and Finland, but only the first mutation would be found in populations that migrated earlier to Siberia or South America. This "genetic clock" shows how additional mutations accumulated over time as more and more of the earth's surface was populated. Still, the number of newer mutations could never

Hi! I'm Gloria!

make up for the way that the initial groups migrating out of Africa were so genetically similar, so humanity has remained quite similar genetically.

By running this genetic clock backwards, the theory that sapiens evolved first in Africa, and then some successfully departed from Africa in small groups around 70,000 years ago, receives additional confirmation. As mentioned already, fewer than 1,000 mating pairs were responsible for the rest of humanity thereafter, both for populations remaining in Africa and for those that departed. Tracking genetic mutations that sons get from their fathers, geneticists can investigate Y-chromosome mutations shared by all male sapiens on the planet. Recent studies asking when those mutations happened point to a date of around 140,000 years ago, which is consistent with archaeological theories about sapiens inhabiting only eastern Africa at that time. Sometimes this theory about the male ancestors of all humans is called the "Y-chromosomal Adam" theory. This label is not accurate, since genetics is not pointing to a "first Adam" human who was the only living man at that time on the planet. Genetics can only indicate that there was a very small population of sapiens at that time, which included one male who first had that mutation.

Mutations can also be traced in genes passed down in mitochondrial DNA, the kind of DNA replicated from mother to offspring—no contribution from fathers is needed. Tracking these mutations similarly reveals how all humans are descended from a very small number of female sapiens living around 200,000 years ago, which is also consistent with the archaeological dating of the origins of Homo sapiens. Although this theory is labeled as the "Mitochondrial Eve" theory, there was no single female giving birth to all subsequent humans. Again, genetics only points to a very small population of sapiens, which included one woman who first had a particular mutation. It is impossible to find in genetics any confirmation for the Biblical tale of Adam and Eve. Besides, the "Y-chromosomal Adam" lived around 60,000 years after "Mitochondrial Eve," so they never could have met, in a garden or anywhere else.

Religion and Racism

Ignorant of our ancestral heritage and our common humanity, the world's religions typically exaggerate superficial traits in order to justify what they perpetuate: prejudice, racism, slavery, and political domination. The earliest civilizations emerged at a time when population sizes had dramatically increased, the density of habitation was much greater, especially where food and water was plentiful, and tribes of the same regions knew each other and could barter together. The Neolithic era had been under way for three thousand years, simple agriculture had begun in the most fertile areas, and settlements along rivers had grown into small villages by 5000 BCE from the Nile and the Euphrates to the Indus of India and the Yellow River of China. Clans had allied together to form tribes of thousands of individuals, and tribes began to forge unstable alliances as minor kingdoms.

The Stone Age worship of heroic tribal ancestors was challenged by the obvious fact that neighboring tribes had quite different ancestral gods. Myths surviving from precivilization tribes involve a variety of gods, each responsible for creating some part of nature, or setting down the pattern for rites and customs. This Stone Age polytheism was transformed where the first civilizations arose and the founding tribes competed for power. The first stage of "civilized" religion involved elevating one's own tribal gods over those of rivals—each tribe proclaiming how their own gods are mightier than those of other tribes. The best practical confirmation of the might of one's feisty gods was apparently victory on the battlefield, so the reputation of one's gods went up or down depending on whether one's tribe held political power.

The religions of early civilizations also featured many gods because their kingdoms suffered from political instability, as centers of power shifted from city to city. So long as one tribal dynasty could rule for awhile, its primary god was credited with creating the cosmos and engendering the other "defeated" gods. When a different tribe of another city rose to power, the god of that city got promoted to supremacy, and so on. After the first four civilizations (the Egyptian, Sumerian, Indus, and Chinese) had lasted a thousand years, their religions were highly polytheistic, because the lists of gods invoked during power struggles had accumulated so many names. These gods were thought to have had a common origin with the cosmos, and whichever god held supremacy carried the responsibility for producing the other gods. Furthermore, the stories about the relationships among the gods were composed in order to legitimize the tribe holding political power. In some myths, the god of a defeated tribe is described as a weak god rightfully defeated by another stronger god. In other myths, the superior god is credited with generating offspring, so that a god of a subordinate tribe is a son or a daughter of the superior god, implying duties of submission and obedience. The way that myths are written and rewritten exposes how the ways of heaven are used for justifying the ways of life on earth.

When the earliest recorded religious stories were written down in the first systems of writing, they were already polytheistic, genealogical, and combative. These "original" myths had little original about them. They already displayed an impressive complexity, due to the dramas of political events over hundreds or thousands of years. Stories about the creation of the first humans got wrapped up in the cosmological versions of these dramas. By the opening of the Bronze Age, around 3000 BCE, religions of civilizations were confident that one or two of the gods were responsible for creating humanity, although they disagreed among each other whether the supreme god(s) had made humans, or lesser gods had gotten around to making humans later on in the course of creation. The Cairo "Hymn to Amun-Re" from the early New Kingdom circa 1500 to 1400 BCE says, "Atum, who made the common man, who distinguished their forms, who made their lives, separated the races (colors) one from another..." Around the same time, the Babylonians elaborated on an older Sumerian tale about the gods creating humans to do the agricultural work. The Babylonians composed a mythological tribute to the great god Marduk, who first settles cosmic matters with divine battles and then later ordered the creation of different types of people to labor in the irrigation fields and serve the rulers of Mesopotamia.

It didn't take much longer for ethnic strife and empire struggles to find an ally in religion, which was quite ready to agree that some people were "more human" than others. By 1000 BCE, a few Hindu myths openly approved of the racial caste system, in which the lighter skinned northerners held power over the darker-skinned southern populations. The Hindu epic *Mahabharata* from around 300 BCE declares that the four primary castes were created by the supreme deity. By then, the

Greek philosopher Aristotle openly stated in his *Politics* a philosophical verdict, something Greeks had long heard in their religion, that peoples from other nations were subhuman barbarians fit only for slavery. Distrust and disdain for other religions, especially those of Egypt, Anatolia (Turkey), the Near East, and Persia continued to be a prominent feature of Western religion after the Greeks. The Romans, the Latin Catholic Church of the "Dark Ages," the medieval Holy Roman Empire, the modern Catholic Church, and Protestantism never wavered from their categorization of Eastern religions as stupidly pagan at best and satanically inspired at worst.

The Roman senator and historian Tacitus said that Egyptians were a "people devoted to superstitions." Despite their official status from Rome as a protected religion, the Roman orator Cicero regarded Jews as "born slaves" and called Judaism a "barbaric superstition, declaring it incompatible with proper Roman order. The Romans commonly regarded Persians as too weak minded and submissive to want liberty or citizenship, capable of understanding only authority and preferring tyrannical kings. The Romans had no difficulty applying these kinds of sentiments against Christian cults spreading through the empire. Once Christianity became the official religion of post-Constantine Rome, it quickly aimed these same prejudices at other religions.

The assessment of Islam by St. Thomas Aquinas, Christianity's foremost medieval thinker and Catholicism's canonical theologian, repeated these kinds of accusations. His verdict on Muhammad:

> He seduced the people by promises of carnal pleasure to which the concupiscence of the flesh goads us. His teachings also contain precepts that were in conformity with his promises, and he gave free rein to carnal pleasure. In all this, as is not unexpected, he was obeyed by carnal men. As for proofs of the truth of his doctrine, he brought forward only such as could be grasped by the natural ability of anyone with a very modest wisdom. Indeed, the truths that he taught he mingled with many fables and with doctrines of the greatest falsity.
> ... What is more, no wise man, men trained in things divine and human, believed in him from the beginning. Those who believed in him were brutal men and desert wanderers, utterly ignorant of all

divine teaching, through whose numbers Mohammed forced others to become his followers by the violence of his arms.[1]

Christianity also inherited an older prejudice against black Africans. The Jewish book of Genesis includes a tale in Chapter 9 about Noah cursing the descendants of his son Ham, who became the people of Canaan. The insertion of this tale about Noah's curse against Canaan was probably an excuse for the Israelite domination of Canaan. However, the Curse of Ham, as this tale was called, specifically said that the people of Canaan/Ham must be servants. Later Talmud commentaries interpreted the meaning of "Ham" according to a similar word for "dark" to explain the origin of people with dark skin. Jews were not enslaving Africans then, but Arabians had been enslaving Africans from across the Nile for a long time. Islam quickly merged these notions. Since much of Genesis is echoed in the Qur'an and Noah is revered by Islam as an early prophet, Islam justified slavery with Noah's supposed curse against dark-skinned peoples. As soon as the European slave trade got under way in the sixteenth century, Christianity revived such ideas, combining them with other Biblical passages approving slavery (for example, Paul's First Letter to the Corinthians and his Letter to Philemon). Until the Civil War, many ministers in the South, and a few in the North, continued to argue that the Bible could not support antislavery efforts. Long after the Civil War, the legend of this "Curse of Ham" continued to fester among racists.

Once a folk tale has momentum, it can easily propagate on its own among people. That's the power of language at work. Language is also the power of people.

The Origin of Language

It's not an exaggeration to say that it is language which makes us human. Religions have never overlooked the core importance of language for humanity. This is a self-interested evaluation for language, of course, since religious stories exist only because we use language. All the same, many myths acknowledge the importance of language for everything else that humans can do.

Indigenous peoples in the South American Andes tell tales about primitive prehuman creatures that could not speak, who had to be destroyed by a good god before the first humans were created. The Wiyot tribe of northern California told a story about how the creator god didn't approve of the first people, who were too furry and couldn't speak correctly. Many more such examples could be listed. Stories from numerous cultures relate how a caring god taught the first humans to speak, and next instructed them in additional skills. Sumerian myth indicates that at first there was one common language, but the great god Enki, who had saved some humans from a great flood, introduced many languages. In Genesis, Moses is told by Yahweh that he taught people how to speak. Where did the different languages come from? After Noah's Flood, the people are said to have tried to build a tower reaching up towards heaven, but God made them abandon the project by giving them different languages.

If there ever was an original language, very little could be known about it. There are no words common to all languages, and languages display diverse kinds of grammar. The secrets to the development of language are hidden in that evolutionary transition from Homo erectus to heidelbergensis and then to sapiens. Communication performed by other species of primates is a suggestive guide to the capacities of erectus. Observed behaviors in primates such as posturing and gesturing can be meaningfully symbolic, as are cries, calls, and barks, along with more intimate noises like grunts, growls, and lip smacking. Because erectus had far larger brains, with extra cortical areas which helped in understanding others, erectus may have been able to use additional symbolic noises, such as singing or whistling, to convey meanings.

However, erectus probably did not have language in the full sense, because its voice box was still small and too high in the throat for making a large range of sounds. With only slight shades of difference between consonants and a limited selection of vowels, erectus could not form many words. Without enough words, communication is very difficult. There's no point to stringing together several words into sentences, because there aren't enough words, and so there's no way to

For Pete's sake, Larry, just SAY it!

form a flexible grammar. Grammar controls word order so that a set of words can be spoken in one sequence to mean something, and that same set of words can be ordered in a different way to mean something quite different. Without grammar, a string of words like "see Jeb tree fall" doesn't have a single meaning, and confusion grows. You and I notice a big difference between "see tree Jeb fall" and "Jeb see tree fall" because our language embodies a grammar and we automatically fill in the extra needed words that would make a complete and fully meaningful sentence. "See tree Jeb fall" could mean "See the tree where Jeb is falling" or maybe "See the tree that Jeb is making fall down." "Jeb see tree fall" might mean "Jeb sees the tree falling down" or maybe "Jeb, see how that tree is falling down." Without a large vocabulary of words or a grammar, language in the full sense could not get started, so erectus was using only a little communication. The same limitation applies to a sign language. Erectus was probably not using a full sign language, but only a small number of gestures to handle immediate needs.

A full language gives speakers the capacity to talk about what has happened in the past, what is happening beyond the visible present, and what may happen in the future. A full language permits people to share and recall instructions about what should be done to accomplish tasks, so that a person can listen to instructions today and complete the task tomorrow. A full language permits speakers to accumulate information over time in their memories, and transmit that collected information to another speaker at any convenient time. Without a full language, people can only learn about what they can directly see. This still permits a small amount of accumulated learning, since an elder can demonstrate accumulated skills before learners who then imitate her. But she has no way of quickly and efficiently explaining all of her skills to a granddaughter. If she doesn't happen to perform some of her acquired skills before a younger audience, those skills die with her.

If erectus did not have full language, then the skills of individual members of erectus couldn't develop much across many generations. Even after hundreds of generations, technical skills wouldn't have developed far. And that is what is seen in the archaeological record of materials left by erectus. The most valuable tools during the Stone Age were made of stone, obviously, but they weren't just any rocks. The Acheulean phase of stone tool making, dating from 1.8 million to around 200,000 years ago, was distinguished by the production of larger, symmetrical, and bifaced hand axes, cleavers, and flakes. These improved stone tools required several stages of careful shaping for achieving the same size every time and maximizing both strength and sharpness. Yet there are few interesting differences between erectus tools across that immense time span. During this span, the brain size of erectus continued to increase, so that extra intelligence was not getting applied to tool development.

Erectus was apparently devoting its technological intelligence to the careful imitation of standard techniques. This achievement must be given proper respect. Even today, months of imitation and practice are required to make a good Acheulean-style tool. However, the evidence suggests that any inventive erectus was unable or unmotivated to teach a new or unusual technique to another erectus. Only later heidelbergensis and neanderthalensis proved to have the capacity to make sophisticated stone tools such as impressive long knives, scrapers for preparing hides, awls for making holes, and spear tips for better hunting. They also started making some stone artifacts that appear to have been far more than just tools, because they display exaggerated features, careful forms, and delicate symmetries making it unlikely that they were used for anything practical. The capacity for describing features of stone and explaining techniques to enhance what can be learned just by observation must have been growing among earlier Homo groups.

The emergence of sapiens coincided with an explosion of new tools from new materials put to novel uses, such as the bow and arrow tipped with small stone points. The technologies of sapiens during the period of 150,000 to 15,000 years ago was remarkably complex. Tools were designed for making other tools, and engineered tools were constructed out of several parts, each requiring many stages of production. Sapiens was still living in the Stone Age, but transportable shelters, bone needles, warm clothing, containers for carrying and storing food, specialized hunting weapons, and fishhooks were all signs that sapiens had reached a high level of technological achievement. Most significantly, that technological sophistication clearly implies that this extensive practical knowledge was not the exclusive possession of single individuals passing skills along only to their watchful children. Groups of skilled practitioners, each specializing in an area of expertise, must have used a full language in order to coordinate, teach, and improve all those technologies. Other core human abilities required a full language too, such as managing larger social groups like clans and tribes, the preservation of wisdom and history in storytelling, and artistic narratives and rituals relating ideas about the world. Without full language, nothing like civilization would have ever developed.

Sapiens was relying on intense communication and instruction to survive in Africa, and the final evolutionary modifications to its brain and throat, marking sapiens (and perhaps neanderthalensis) off from heidelbergensis, are distinctive features. The lower position of the larynx in the throat and the repositioning of the tongue permitted sapiens to make the complete range of phonetic sounds that we can hear today. The parts of the cortex responsible for making and processing vocal sounds attained their full development. Broca's area in the left hemisphere, an area helping with understanding complex sentence structure, had grown larger, for example. Neanderthalensis may not have been far behind in their acquisition of full language, but if language was acquired, it wasn't put to use for developing better tools. Perhaps later neanderthalensis groups were under too much pressure to barely survive, and became extinct too soon to take full advantage of language.

Languages which did survive to the present day display family relationships that correspond to the paths of human migration across the continents. Indigenous languages in North and South America have similarities among each other, and they all are related to languages in Asia. These relationships make sense because the Americas were first populated by peoples from northern and southern Asia. Most languages from Europe, Russia, Turkey, Iran, and India are related to an ancestor language spoken in Anatolia around 9,000 BCE. Variations on that language spread west and east, as agricultural ideas radiated outwards from the Levant. The languages of the people who knew how to practice agriculture were adopted by a neighboring region as they acquired agriculture in turn. Languages tended to follow cultural innovations as they were adapted by neighboring lands. In our own time, English is the language that many educated people around the world must learn in order to keep up with scientific and technological progress.

Notes

1. Aquinas, Thomas. *Summa Contra Gentiles*, trans. A.C. Pegis (Notre Dame, Ind.: University of Notre Dame Press, 1975), pp. 73–74.

Further Reading

Allan, Tony, Fergus Fleming, and Charles Phillips. *African Myths and Beliefs*. New York: Rosen Publishing Group, 2012.

Beck Roger B. *The History of South Africa*. Westport, Conn.: Greenwood, 2000.

Crawford, Suzanne J., and Dennis F. Kelley, ed. *American Indian Religious Traditions: An Encyclopedia*. Santa Barbara, Cal.: ABC-CLIO, 2005.

Falk, Dean. *The Fossil Chronicles: How Two Controversial Discoveries Changed Our View of Human Evolution*. Berkeley: University of California Press, 2011.

Fitch, W. Tecumseh. *The Evolution of Language*. Cambridge, UK: Cambridge University Press, 2010.

Fluehr-Lobban, Carolyn. *Race and Racism: An Introduction*. Lanham, Md.: Rowman Altamira, 2006.

Harris, Eugene E. *Ancestors in Our Genome: The New Science of Human Evolution*. Oxford: Oxford University Press, 2014.

Isaac, Benjamin. *The Invention of Racism in Classical Antiquity*. Princeton, N.J.: Princeton University Press, 2013.

Leeming, David. *Creation Myths of the World: An Encyclopedia*, 2nd edn. Santa Barbara, Cal.: ABC-CLIO, 2010.

Masataka, Nobuo, ed. *The Origins of Language: Unraveling Evolutionary Forces*. Berlin and New York: Springer, 2008.

Meredith, Martin. *Born in Africa: The Quest for the Origins of Human Life*. New York: PublicAffairs, 2012.

Ostler, Nicholas. *Empires of the Word: A Language History of the World*. New York: HarperCollins, 2011.

Patel, Aniruddh D. *Music, Language, and the Brain*. Oxford: Oxford University Press, 2010.

Pengzhi, Lü, and John Lagerwey, ed. *Early Chinese Religion, Part One: Shang through Han.* Leiden and Boston: Brill, 2010.

Penna, Anthony N. *The Human Footprint: A Global Environmental History.* Malden, Mass.: Wiley-Blackwell, 2010

Pinch, Geraldine. *Egyptian Mythology: A Guide to the Gods, Goddesses, and Traditions of Ancient Egypt.* Oxford: Oxford University Press, 2003.

Reynolds, Sally C., and Andrew Gallagher, ed. *African Genesis: Perspectives on Hominid Evolution.* Cambridge, UK: Cambridge University Press, 2012.

Spier, Fred. *Big History and the Future of Humanity.* Malden, Mass.: Wiley-Blackwell, 2011.

Staller, John E., ed. *Pre-Columbian Landscapes of Creation and Origin.* Berlin and New York: Springer, 2008.

Tattersall, Ian. *The World from Beginnings to 4000 BCE.* Oxford: Oxford University Press, 2008.

Teel, Karen. *Racism and the Image of God.* London: Palgrave Macmillan, 2010.

Walter, Chip. *Last Ape Standing: The Seven-Million-Year Story of How and Why We Survived.* London: Bloomsbury, 2013.

Willoughby, Pamela R. *The Evolution of Modern Humans in Africa: A Comprehensive Guide.* Lanham, Md.: AltaMira Press, 2007.

5

HUMAN CULTURE

Religions tend to mirror the diversity of human society. The way that religion can flexibly function to support the local social order, whatever that order is like, explains why religions disagree over many small things about life. Cultural anthropology and sociology provide insights into cultural diversity, but nature v. nurture debates remain: what is innate and the same for all humans, and what is entirely conventional? From religion's perspective, the question is more about this: what is desired from humans by the Creator, and what is left up to us?

When we think about culture, what may first come to mind is material culture: all the tools and technologies, all the things assembled and constructed, which make our existence both easier and busier. Religions might like to make it appear that they can reside in a parallel world alongside material culture. That distance is illusory. Religions are happy to use technologies as material means when they prove useful to promoting religion's immaterial ends. Religions are also capable of disapproving technologies for disrupting the social order too much or deviating from theological dictates. Extremely conservative religions may reject new technologies, such as machines, threatening a simpler way of life. Dogmatic theologies may reject new scientific instruments, such as the telescope, for their danger to sacred worldviews of the past. Religions generally monitor material culture for threats against what religion really cares about within culture: proper social order and strict morality, and indoctrinating a worldview reinforcing them. The human realm of character and conduct is religion's primary focus, as far as this mortal life goes.

Social Roles and Rules

All human cultures use social roles to assign powers, responsibilities, and specialized tasks to different people. Religions do not invent basic social roles, nor do they invent the moral codes that dictate what counts as virtuous fulfillment of social responsibilities. Morality is far older than any religion. From the origins of Homo sapiens, and possibly deeper in Homo ancestry, humans have used a basic universal morality about avoiding harms, conflicts, and distrusts that obstruct helpful cooperation.

As the first religious ideas arose, sometime between 100,000 and 50,000 years ago, these ideas emerged within cultural environments expecting conformity to not just that basic morality, but also to distinctive social relationships. The relationships between men and women, parents and children, members of extended kinship families, and members of a clan under authority figures were already embedded in Homo sapiens. Religion did not establish who mothers and fathers were, or originate gender roles, or set down who should be in charge of group decisions, although religions have manipulated those features of social life.

By the time the earliest recorded religious narratives were composed by priestly writers in the first civilizations, societies had grown amazingly complex, requiring more social roles and duties than could be listed in any tale. Religious texts identified the most important roles of their day, from the king and the various offices that kept the government working, down to the many kinds of laboring classes such as farmers and stone cutters. And each religion stated that it knew how and why those social roles had been invented. Ordinary people using their own thinking were not responsible for planning out society. Only the gods could have done it.

Not sure I like the sound of that...

Without religion's stories about godly deeds and decrees, no one would know their true and rightful place in the order of things, so no one would understand their duties, and the gods would be greatly displeased. Religions of every human society have said that they alone know why society should run the way it does, and why the god(s) require society to never deviate from that way. Religion is inherently conservative: the origins of everything long ago explain why today's society should return to that original plan. To be truly human in the way you were designed, you must conform to social expectations, because those expectations are divine. Your destiny supposedly depends on it.

Religions enforce social roles through appeals to morality: people are told to do what their social roles require, as parents or children, masters or servants (etc.), in order to be fully virtuous. To assign different responsibilities to different roles, those roles must be attached to distinct moral codes. If there were one moral code applying to everyone, no grounds for social roles could be found in morality, and everyone would be socially equal in the eyes of morality. But religions everywhere enforce social roles, and enforce them with all the powers they have. Religions are able to tell one set of people about one moral code, and another set of people about a different moral code, and justify these two separate moral codes with stories about the creation of humanity by gods who don't view people equally. Everyone can be virtuous, so long as they obey their assigned social roles, roles approved by the creator(s).

A religion that happens to be dominant in a society is dominant because it endorses that society's structure and social roles. A religion that disagrees with much of a society's way of doing things is always in the minority, and probably arose in a neighboring society. When religious dissent arises within a society, it will either soon die out, migrate to another more hospitable society, or force society to conform to it (this rarely happens). In the long run, there is always a close match between a dominant religion and a society's structure, rather than a deep antagonism. This is easily observed to be the most common relationship discovered around the world between culture and religion: they are made for each other and support each other. In small societies, it is impossible to tell exactly where religion stops and wider culture begins, because they infuse and empower each other so thoroughly. In very large and ethnically diverse societies, such as empires and large countries, minority religions that can't agree with the social majority may be tolerated (but usually not), and they may share in political power (but usually not, except in highly democratic countries). All this evidence points to the way that religions are designed to mirror the cultures of the societies they dominate, but they dominate because they follow that culture. Religions don't invent social structures or cultures, but they do reinforce them, and help justify cultural ways to the people.

When a religion has to deal with claims that it wrongly supports a harmful rule of a social role, it can dodge those accusations with several diversion tactics. One diversion is to say that such a harmful rule isn't in scripture. Another is to say that most people don't do it, or that religious leaders don't endorse it. Yet another is to point to the way that many followers of that religion have disagreed with that rule. Still another is to claim that religion has had nothing to do with that rule, since it is just a "cultural" matter instead.

Scriptures can be a very convenient resource. When accused of supporting a harmful rule, a religion can appeal to its own scripture, pointing out how there's no place in scripture where that specific rule is endorsed. But exact scripture is hardly needed to religiously justify social roles. Where paradigm examples are easily read in scripture, only a little thought is needed to extend examples further. If the Bible says that Eve was made from Adam to help him, and Eve could be tricked by a talking snake, and Eve and all women deserve a peculiar divine punishment, and women must be silent in church, and so on and on, all sorts of additional rules keeping women subordinate to men can be inferred (by men). Nowhere in the Bible does it precisely say that women should be property, or that people of darker skin should be property, yet the Bible was quite sufficient to enforce such roles, and dominant Christianity accepted that social order. A religion doesn't have to directly dictate specific roles or practices in order to support them or just quietly accept them.

Nor does an entire religion have to require an objectionable practice. Even if a small minority of followers ever do it, the way that a religion as a whole fails to condemn it reveals how that religion condones that practice. If some religious leaders ignore it, or even encourage resistance to it, that still leaves the religion in general responsible for permitting it, so it can't simply be a "cultural" matter. Even where an objectionable cultural practice is older than the current dominant religion, so that this religion simply arrived and accepted that practice without comment, that religion remains responsible if that practice is allowed to flourish unopposed.

What is natural? Not surprisingly, every culture regards its own social order and civic practices as entirely natural. Religions reinforce their cultures, so religions similarly regard local folkways as a reliable guide to what is natural for all humanity. Religions that spread throughout neighboring societies carry the folkways from the society of origin. A few large religions have spread throughout so many societies that they stop trying to enforce just one culture, instead requiring a smaller number of distinctive rules that people of many cultures can all follow. The way that Christianity had to leave behind almost all of the Jewish rules, preserving only male circumcision and a few rituals, is one example; Buddhism's distillation into meditative practices as it went east to China and Japan (and on to North America), leaving behind Indian culture, is another example.

One might suppose that the ability of large religions to tolerate a high degree of cultural diversity would make their religious followers less hostile to ways of other societies and people of other religions. The exact opposite remains the case. Religion reinforces convictions that one's own core folkways are the best ways to live. We aren't talking only about the kinds of foods eaten where a person lives, or the style of architecture preferred. Core folkways include rules, spoken and unspoken, about how families work, how children are raised, how one becomes an adult, what jobs are appropriate, how one treats one's neighbors, and how communities handle good times and bad. In one society, physical punishment of children or letting children drink alcohol is quite natural, while in another society they are evils. In one society, the elderly should live with their children; in another society the elderly try to stay independent. The same broad religion can staunchly agree with both societies on their separate territories without ever feeling like a contradiction is involved. Large religions like Christianity, Islam, Hinduism, and Buddhism developed highly flexible theologies to permit them to adapt to different societies. Polytheistic religions adopt local gods and their local ways, explaining that local gods are either manifestations of the one God or they are part of the one God's family. Monotheistic religions keep up beliefs in spiritual powers by calling them angels or saints; local gods get converted into angels or saints as the larger religion gains converts, so that divine authorization of local folkways can continue undisturbed.

What religion evidently does for each follower is to reassure them that God approves of their core folkways, and followers are loyal to that religion because they feel reassured. A side effect of this confidence in local ways is a xenophobic distrust and disdain towards other peoples. Religious belief is strongly correlated with mutual approval and trust among one's local religious friends, and disapproval or even hatred against strangers from other societies. The social sciences have long been studying how religion increases trust among congregants who already know each other inside a church, but trust drops quickly towards unfamiliar people. If the cause for this decrease in trust is just the prior social familiarity, and not any religious factor, then all religious groups and nonreligious groups would roughly display the same pattern, yet they do not. Conservative Protestants are more distrusting of strangers than nonreligious groups or Mainline Protestants, and Catholics are even more distrusting. Clearly, conservative religion fails to "love thy neighbor" unless thy neighbor is already in the same church. The effects of religion's inability to decrease fear of strangers extends beyond church walls, too. People of any religion or no religion simply standing near churches display less trust and more hostility towards all sorts of social groups. How much a religion preaches love towards all doesn't have any significant influence. Despite all the preaching, religion and xenophobia go together.

Religious people are generally thought to be more dogmatic and conservative. That stereotype generally happens to be correct. Researchers have long been investigating relationships between strong religious conviction and psychological traits. For example, are rigidly religious people more likely to be people who are prejudiced, authoritarian, or militaristic? Research has confirmed positive correlations between the strength of religious conviction and unfortunate traits such as authoritarianism and ethnocentrism.[1] Studies have compared "fundamentalist" conviction with the "questing" style of religious belief—fundamentalists think their religion supplies infallible truths while questers just want personal spiritual growth. Fundamentalism is correlated positively with authoritarianism, militarism, and prejudice, whereas questing is correlated with less authoritarianism and less prejudice. Additional strong correlations are found between fundamentalism and ethnocentrism, prejudiced attitudes against "deviant" lifestyles, intellectual rigidity, a preference for stern authority, and a higher willingness to impose just one narrow way of life on everyone else, by force if necessary. The long bloody history of religious intolerance and religious war is not accidental or just a product of ignorant bygone times.

Women and Family
Religions claim to know the true nature and proper place of people in society. Without any knowledge about Homo sapiens from consulting nature or scientific inquiry, religions feel entitled to say exactly who and what a person is, and what they are destined to be. The narrow categories into which everyone must fit serve to constrict and control what people can do with their lives. Religion only superficially concerns the next life—it is obsessed with this life, and with each person's life. It is so concerned for proper conduct and correct social roles that it overrides any rival ideas about how to live this life and it punishes deviance with earthly suffering here in this world and terrible suffering in the next world. Using its "knowledge" of the world's Creator and what this Creator loves, religions teach their followers to hate anything that can't live up to divine expectations.

The safest general rule about religion and the family is that religions enforce monogamy and socially sanctioned marriage, except for all the religions that don't. That is to say, religions notoriously follow local cultural custom. Religion didn't invent the idea of monogamy and it had nothing to do with inventing the social enforcement of marriage. There have been some religions that care little for monogamy and tolerate many kinds of partnerships. Where societies recognize marriage and enforce rules about marriage, the religions of those societies sternly urge obedience to exactly those rules because divine powers approve of just those rules. And another society's different rules are likewise rigidly reinforced by its religion. If a new religion wants to enforce rules about sexual partnerships that differ from the surrounding society, it must establish a new society, because it will not be absorbed into the old. Religions interested in celibacy, polygamy, or communal sexuality, for example, soon find that starting separate small communes or societies is necessary for survival. Catholic monks or Shakers practicing celibacy are one example, while Mormons founding Utah and defending polygamy is another.

You're saying we should only have **how many** *wives... ?*

It must be recognized that most of the world's major religions have agreed that monogamy between heterosexual partners for the purpose of childrearing is the ideal form of family life. It also must be recognized that family life is much more controlled by men than by women in those religions. Although religions regard family life as "natural" for men, societies generally allow men many more options about the timing and arrangements of marriage and family status, including ending a partnership. When it comes to women, religions permit far fewer options.

Hinduism still includes the practice of "sati"—when the wife kills herself after her husband's death—despite repeated efforts by the British government and the independent Indian government to eliminate it. It still occurs every year in India, and when it does happen, the outrage among modern Indians is matched by vehement defenses of sati from cross sections of Hinduism, especially conservative sects and certain castes. The defenses are always the same: only a small number of wives commit sati nowadays; the government should stay out of religious expression; these women are saintly examples of female virtue; and some Hindu scriptures condone this practice. The term "sati" is revealing, since it roughly means "what woman truly is" and that is understood by Hinduism to mean that a woman truly exists only through her relationship with her husband. Exemplifying genuine sati practically means being wherever her husband is—if she is somewhere else, she has lost her own existence—therefore, if her husband is dead, then she must be dead also, to remain a true woman. The fact that only a minority of wives committed sati over the centuries doesn't erase the fact that traditional Hinduism has always regarded women as only fit for joining men and following men. The practice of sati, and the staunch defense of sati, exemplifies Hinduism's firm convictions about the proper subservient role of women. That is why criticism of sati arouses vast outrage from many traditional Hindus.

Most religions include creation stories or legends of male heroism explaining why men are superior to women and deserve more powerful roles in society than women. Because religions operate in the realm of moral reinforcement, they typically explain that women must display special female virtues no less important than male virtues. Some religions don't bother to make it seem like women have special virtues to justify their separate status. The Masai of eastern Africa have a legend about women which says that they used to be more like warriors than men, but long ago the men rebelled and attacked the women, scarring them with vaginas and leaving them only fit for childbearing and family duties. Fascination with female sexual organs, and endless rites and rules controlling those organs, are so common among religions that it is difficult to find any religions that ignore them. The time of first menstruation, monthly menstruation, the pregnancy term, and the event of childbirth are full of religious significance and momentous concern. Typical among the rites and rules, found among indigenous peoples from the Native Americans to southeast Asia and Polynesia, are specific rules restricting contact with a menstruating or pregnant woman. Other religious restrictions confining a pregnant women to hearth and home, and keeping her at home to raise children, are also too many to enumerate.

Religions all too commonly reinforce what many cultures require: viewing females as holding special moral duties of womanhood and motherhood, duties based on some divine regard for the female that must be imitated on earth. Holding women in special regard due to a privileged status in the eyes of a god can only mean, at least to most religions, that women cannot undertake the male burdens of getting educated, managing property, participating in civic affairs, and holding offices of religious leadership. There are notable exceptions among the world's religions where women have about as much opportunity to hold social positions of prestige and power as men, but these exceptions prove the rule: religion in general has helped ensure the perpetuation of male domination. Another typical feature of religions is the expectation that only men should be religious mystics, thinkers, leaders, and saints. Even where a religion allows women to have separate institutions to display great devotion, those institutions are smaller, not well funded, modeled after male institutions, given far less social prestige, and ruled by men.

No matter how much metaphysical importance is assigned to the female in a religion's cosmological scheme, women end up with different and inferior social roles and less access to equal social status. Chinese religions such as Daoism and Confucianism treat the ultimate cosmic powers as no less female than male, and portray balance and harmony between them as an ideal for the heavens as well as earth. Nonetheless, a succession of Chinese dynasties and empires from 2000 BCE down to the twentieth century haven't promoted much female equality to match such an ideal. After Buddhism spread across eastern Asia, from Korea and China south to Thailand, it tended to support the patriarchal cultures already there before it. Early Buddhism may have had egalitarian and gender-equality phases, but its origins and its migrations beyond India betrayed those ideals. In India, the record of male domination over women is unrelenting. The Laws of Manu, the most important codification of Hinduism dating from around 100 CE, doesn't fail to put women in a small and subordinate place. The texts numbered 5.147–149 say: "A girl, a young woman, or even an old woman should not do anything independently, even in (her own) house. In childhood a woman should be under her father's control, in youth under her husband's, and when her husband is dead, under her sons'. She should not have independence. A woman should not try to separate herself from her father, her husband, or her sons, for her separation from them would make both (her own and her husband's) families contemptible."[2]

Matters are even worse in religions that either elevate male gods over the females goddesses, or worship just one male god. The Jewish law in The Mishnah, Tractate Ketubot 5:5 and 5:8, says: "These are the tasks that a wife carries out for her husband: grinding corn, baking, washing, cooking, suckling her child, making his bed for him, and working in wool. If she brings with her one maidservant [into the marriage], she need not grind, bake, or wash; [if she bring in] two, she need not cook, nor suckle her child; three [maidservants], she need not make his bed, nor work in wool; four [maidservants], she may sit on a high seat [i.e., not work at all]. Rabbi Eliezer says: even if she brought into the marriage one hundred maidservants, he may compel her to work in wool, for idleness leads to lewdness."[3]

Christianity and Islam followed the Jewish example, and Middle Eastern cultures in general, by carefully restricting a woman's life to exist in subservience to men. Women were destined for marriage to men, and consigned to the care of children and home life. Like Judaism, Christianity and Islam strongly disapprove of sex outside of marriage, and find ways to punish and obstruct women more heavily than men for premarital sex, infidelity during marriage, seeking a divorce, or getting remarried. There are numerous passages in the New Testament where support for paternalistic control of men over women can easily be found, such as this King James verse in Colossians 3:18—"Wives, submit yourselves unto your own husbands, as it is fit in the Lord." The Qur'an includes passages that clearly subordinate women to men, such as verse 4:34: "Men are the protectors and maintainers of women, because God has given the one more (strength) than the other and because they support them from their means."[4] Later legends about Muhammad's life, commentaries on the Qur'an, and the development of Islamic law expanded on the subordinate role of women. From child marriage and polygamy to restrictive divorce laws and the control of children by husbands, Islam has proven to be one of most antifemale religions on record. Islam didn't have to invent most of these rules, since Islamic law usually follows much older custom wherever it has flourished, and those cultures have been overwhelmingly patriarchal.

Sexuality, Mental Illness, and "Deviant" Behaviors

Deviations from "normality" receive extraordinary scrutiny by religions. A small number of religions across history and around the world today make a respected place for homosexuality and a protected space for the mentally ill. That small number only serves to illustrate how most religions, especially the religions of large empires and entire civilizations, have regarded deviancy as not just immoral but also as signs of sinful depravity in the eyes of god.

Until near the close of the twentieth century, homosexuality and other nonhetero sexual statuses were grounds for second-class citizenship and punishable criminal offenses in many prominent countries, including the United States, United Kingdom, Germany, Australia, China, Russia, South Africa, and Chile. By the start of the twenty-first century, almost all of the countries dominated by Christianity had repealed laws criminalizing nonhetero sexuality, and many had started to recognize equal civil liberties, such as marriage and adoption, for nonhetero sexual persons. Today, most of the countries still criminalizing "deviant" sexualities are located in Africa, the Middle East, and Central Asia, including almost all Islam-dominated countries. Homosexuality is punishable by death in Sudan, Saudi Arabia, and Iran. Even where homosexuality is no longer criminalized, nonhetero sexual persons lack full equal civil rights and civil liberties except in a handful of European countries, and hateful intolerance against alternative sexualities is still widely practiced and rarely punished.

Religious prejudice against nonhetero sexual practices is almost entirely responsible for justifying this degree of discrimination in peoples' minds. In Christian-dominated countries, several large denominations have denounced this prejudice and accepted all people equally, providing public leadership on issues such as marriage. But other large conservative denominations, such as the Roman Catholic Church and Southern Baptists, resist progress. They continue to classify nonhetero sexual status as specifically forbidden in the Bible by God as a grave sin and hence punishable with eternal damnation in Hell.

Religions in general feel perfectly at liberty to classify "deviant" people as rebelliously unworthy and sinful in the eyes of the Creator of the world. They aren't interested in actual information about the world along the way, so presenting scientific evidence about the naturality and natural explanations for many varieties of hetero and nonhetero status and practices seems to have little effect on religious dogmatism. The naturality of all varieties of sexuality should tell any reasonable person that religion is incompetent to say who shall be regarded as "normal" and "right." Religions are quite capable of perpetuating hateful prejudices based on complete ignorance about humanity and human nature, but they have few resources for reversing cultural inertia and challenging popular prejudice.

This idea is just decades ahead of its time!

Another example of this religious incapacity to deal with scientific information about human beings is the terrible record of religion on mental illnesses and disabilities. Christianity has traditionally regarded mental illness as either a sign of personal sinfulness, a sign of demonic possession by an evil power, or as a punishment for sinful deeds. The Bible isn't the clear source of this prejudice—mental illnesses aren't the subject of divine decrees and Jesus heals some people from their mental problems without labeling them as sinful. Nevertheless, popular Christianity has always taken mental illness and mental disability to be signs of sinfulness and God's disfavor, and almost all theologians of the Church, Catholic and Protestant, have confirmed that prejudice. Not until the late nineteenth century did many church leaders relent from their position that there's no relevant difference between freely chosen sinful behavior and abnormal behavior due to neurological disturbances or diseases. The Catholic tradition, including the Roman, Greek, and Russian branches, found mental disturbances to be an opportunity for diagnosing sin or demonic possession, and Roman

Catholicism continues to practice exorcisms against demons down to this day.

Evangelical Protestantism from its beginnings also resisted the Enlightenment idea that diseases in general, and mental disturbances in particular, are only evidence of physiological disorders and not signs of a person's alliance with the forces of Satan. The greatest obstacles to establishing the medical profession's ability to care for the insane and mentally ill came from ministers who declared that treatments for the mentally deviant are against God's will. Using their political influence, some conservative Protestant ministers tried to prevent the establishment of insane asylums and institutions for the mentally disabled. Having already declared religion to be the genuine way to understand all things having to do with the mind and the soul, these religious leaders were extremely hostile to medicine's intrusion on their rightful territory of "expertise." Other conservative Christians were happy to use mental illness institutions to house men and women (mostly women) who were not religiously devout, or became freethinkers against the church. Liberal Christianity relented during the nineteenth century and conservative religion retreated from the territory of medicine during the twentieth century. Smaller religious denominations continue to categorize the mentally ill and disabled as sinfully lost, including Christian Science, Jehovah's Witnesses, and Pentecostals.

Criminal conduct by mentally sound people has also been long classified by religions as sinful disregard for God's laws. Religions haven't been humble about their "knowledge" of the causes and cures for crime. Religions typically regard themselves as the best authorities over what should count as crime. Only with modern, secular governments can criminality be separated from sinfulness, so that churches are no longer responsible for determining the laws of a nation. Where governments are not secular, religious law is not easily distinguishable from civil law, and religious leaders have a much greater role deciding what shall count as criminal and how society shall deal with criminals. Even where the criminal justice system isn't connected to a society's churches, a religion may still influence how criminals are punished and rehabilitated. Although the percentage of nonbelievers in society is higher than the percentage in prisons, popular religious ideas still prevail that lack of religion is a large cause of crime, and that religious faith prevents crime. After criminals are in prison, ideas about rehabilitating them for a return to society frequently have a religious aspect. Ministering to criminals has long been regarded as a duty of preachers. Religious people like to suppose that criminals who become more religious in prison are less likely to stay criminals after prison, despite the lack of supporting evidence.[5]

Education

In a sense, culture rests on education, since little knowledge can be passed down to the next generations without careful instruction. After the invention of writing around 3000 BCE, this powerful and dangerous technology was kept under tight bureaucratic control. Only priests and scribes could use writing, a technical skill acquired from membership in guilds that regulated membership most carefully, that was normally used only for state business under the most strict supervision. The high aristocracies of advanced civilizations also used writing for communicating private matters among family, friends, and allies, usually without fear of being intercepted and understood, since writing among the literate was already in its own impenetrable code of symbols—letters and numbers. By 500 BCE, perhaps only 2% of adult men could read and write in the most cosmopolitan cities like the capitals of Egypt and Persia.

The modern world assumes that the ability to read and write is essential to good public order, sustaining high employment, and maintaining civic virtues. That assumption was new to most of the world in the nineteenth century. Three centuries ago, the exact opposite assumption—that keeping almost all people illiterate is necessary to orderly society—controlled almost the entire world. The only exceptions were Protestant countries having stable governments. Protestant Christianity encouraged the reading of the King James Bible by everyone, and literacy was very high for that era, receiving an additional lift from the Enlightenment after 1750. In Puritan New England, even women could typically read and write, although only religious texts were widely available. During the nineteenth century, literacy rates continued to rise in Protestant countries, explaining why those countries continued to be among the most educated in the world down to the present day. Literacy also started to rise above 30% in revolutionary France after a secular government replaced the "divinely appointed" monarchy, and continued to rise despite the return of the monarchy in the early nineteenth century.

Europe's aristocracy and established Churches were still the greatest obstacle to mass education. Bernard de Mandeville (1670–1733) was an Enlightenment figure and no friend of monarchy and aristocracy, yet his satirical comments on public education do represent the dominant stance against education in Europe during his day. He wrote,

> To make the society happy, and the people easy under the meanest circumstances, it is requisite that great numbers of them be ignorant as well as poor. ... The welfare and felicity therefore of every state and kingdom, require that the knowledge of the working poor should be confined within the verge of their occupations, and never extended (as to things visible) beyond what relates to their calling. The more a shepherd, a plowman, or any other peasant, knows of the world, and the things that are foreign to his labour or employment, the less fit he will be to go through the fatigues and hardships of it with cheerfulness and content.[6]

Where Catholicism continued to dominate in southern and eastern Europe and in Central and South America, literacy lagged far behind, since it was mostly reserved for the aristocracy, the military, and those entering religious seminaries. Catholicism couldn't even promote the reading of the Bible, refusing for a thousand years to permit translations from the Latin Bible into the common European languages such as Italian, French, or English. Across Europe and in the United States, Catholicism fought bitterly against public school education well into the twentieth century, and in South America, Catholicism successfully obstructed education for the peasant and working classes in alliances with military dictatorships until well after World War II.

Russia embarked on greater schooling for children after the 1917 Revolution towards communism, and soon achieved near-universal literacy. Muslim countries inaugurated the availability of schooling for their populations after the fall of the Caliphate in the wake of World War I. Education was even less accessible in India and China, the only other regions of the world where significant quantities of reading texts were even available. In 1900, the literacy rates in those civilizations were little higher than what it had been in the Roman Empire two millennia earlier, perhaps 10% among men. The occupation of India by the British Empire in the nineteenth century, and the rise of communism in China in the twentieth century, witnessed the slow rise of literacy, although rural poor areas of both countries still have high illiteracy rates. Outside of North America, Europe, Russia, India, and China, literacy was not regarded as a priority until well into the twentieth century. By then, most countries had realized that their progress as civilized and developing nations depended on universal education.

Today, religion is associated with possessing education. Does it really deserve that reputation? During the nineteenth centuries in America, Europe, and the Middle East, religion had to be associated with education because powerful religious forces demanded and received license to control the education of children in religious schools. Very little public and secular education had been established, so anything beyond basic reading and writing had to be obtained from religious schools and almost all teachers were themselves trained, and typically certified or even ordained, by religious institutions. Those religious institutions had inherited a long tradition of higher education and scholarship from the colleges and universities that had flourished from Spain to Iran since medieval times. Universities were tightly controlled by religious authorities across Christendom and the Islamic Empire precisely because education was such an immense threat to the control of ideas by religion. Illiterates could not become intelligent heretics or rivals to bureaucratic power. Those who could read might read unapproved texts, especially the writings of other peoples from other times.

Christian and Islamic universities receive credit for preserving Egyptian, Persian, Greek, and Roman texts. But they are also responsible for systematically destroying most of those civilizations' writings as they got their hands on preserved libraries. Only a tiny fraction of the vast literary and scientific output from earlier civilizations survived that terrible fire of "higher education" in the grip of religious dogmatism and fanaticism. Many writings dating from before 400 CE only survived because the parchment or paper they were written on was too scarce and valuable to be destroyed. Although those older texts got erased or written over, modern techniques can still trace the original lettering. But almost all of the culture and knowledge from earlier civilizations was deliberately and systemically destroyed. Without doubt the greatest enemy of culture and education the world has ever known is religion. Religions fastidiously preserve anything from their own traditions, and venomously eliminate anything different. Religions that forbid icons and depictions of its own gods are especially hostile towards anything representing other gods, and the scriptures and histories of rival religions are always obvious targets.

The world's greatest library, The Library of Alexandria, was allowed to decay during religious upheaval after the Roman conquest, and its remaining contents were destroyed with the arrival of Christianity. One major surviving collection, the Serapeum of Alexandria, was burnt at the order of Archbishop Theophilus in 391 CE, and Hypatia, the pagan scientist, was murdered in Alexandria by Christians in 415 CE. Islam is also linked to the final fate of the Alexandria Library after it arrived in 641 CE, but that connection is doubtful, since the chances that anything had remained under Christian rule were very small. Still, legend preserves a saying from Caliph Umar, father-in-law of Muhammad, who reportedly instructed that any library should be destroyed, since its books either agree with the Qur'an, so they are not needed, or they don't agree, in which case they are disposable. Since such legendary sayings almost have the force of Islamic law, this attitude towards non-Islamic learning does represent the widespread attitude of Muslim peoples down to the present day.

Alexandria, AD 391

Tenochtitlan, AD 1520

Civilization did have a chance under Islam. During the height of the Islamic Empire, its central cities from Cordoba in Spain to Bagdad in Iraq accumulated impressive libraries of both Muslin knowledge and some Greek, Roman, Persian, and Egyptian knowledge. Europe would know nothing of Aristotle were it not for Spanish libraries, looted and translated after Christianity's reconquest during the thirteenth century. The timing was good for a rescue, ironically, since by then Islam was degenerating back into fanaticism and fundamentalist dogmatism—even the great Muslim scientists and philosophers were under attack and their writings destroyed by their fellow Muslims. In eastern parts of the Muslim Empire under the Abbasid Caliphate, contact with Hinduism and the Mongols fanned the flames of religious hatred. Muslim warlord Muhammad Bakhtiyar Khilji extended Islam's reach deep into India after 1200, happily destroying cities along the way, and his casualties included the burning of the Buddhist Nalanda University and the Vikramasila University. Muslims continued to destroy cultural centers and libraries of any non-Muslim culture that they could invade for century after century, continuing down to present times. Many Muslims today cannot even understand the value of their own cultural heritage. The Institute of Egypt, founded in 1798 and one of the greatest museums in Cairo, was burned and destroyed by Muslim mobs on 17 December 2011. Ongoing destruction of non-Muslim centers of culture by Islamic radicals is now occurring in Afghanistan, Pakistan, and central Africa.

India and China do not have a much better record of advancing literacy. Hinduism could not be a positive force for education, since its local priestly leaders, fakirs, and gurus weren't expected to teach anyone how to read, and study of the Hindu scriptures was regarded as the exclusive responsibility of aristocratic male Brahmins. Hinduism was never as eager to destroy non-Hindu culture and literature, instead absorbing foreign ideas into its all-encompassing pantheon of iconography and local gods. The reliance of Hinduism on participatory ritual and visual iconography to inspire spirituality and devotion meant that the written word was left out as irrelevant to religious participation. Buddhism was even less reliant on sacred writings; masters could understand some of the Buddhist texts, but most monks needed only expertise in meditation practices. During medieval periods of Muslim domination over India, women were especially forbidden from obtaining any education, and the restoration of Hindu rule didn't improve the condition of women. Only late into the reign of the British over India did education for the masses receive much attention.

Chinese religion did assist the rise of literacy, since religious thinking was closely connected with proper government administration and a pragmatic understanding of civic life. Basic literacy among men living in urban areas may have been as high as 40% by the 1800s, but only 10% among women. Daoism, Confucianism, and post-Confucian schools of thought produced a very large number of writings about everything from politics, law, and public policy to medicine and agriculture. This practical "library" of information about how to run one's own life and community and how to operate a society or kingdom attracted much interest in the ability to read, and many families in heavily populated regions of China had at least one member who could read well. This does not mean that China always respected learning and scholarship. The "Warring States" period before the unification of China in 221 BCE was an intellectual period of free thinking and inventive philosophizing. The first great emperor of China, Qin Shi Huang (Shih Huang-ti), reportedly ordered the burning of any books having to do with topics such as philosophy, science, politics, history, and religion in 213 BCE in order to eliminate competition with his own legalist justification for absolute monarchy. Later emperors imitated this example, combating any intellectual rivalry by destroying libraries and centers of learning.

Around the world, the only consistent advancement of public literacy and education for everyone has been due to modern secular governments separating church from state. As the long record of history displays, where religion has dominated culture, the ability to think for oneself is not tolerated, and education has only suffered. If religion continues to be dominant in regions of the world where the people remain ignorant, religion itself is to blame.

Virtue, Power, and Violence

The origin of the word "virtue" comes from Roman culture, in which "vir" means "male," and virtue meant masculine traits. Masculine traits for the Romans should sound familiar. Men should be in control, including control over themselves and control over their families and family finances. That powerful control in turn requires the masculine traits of rationality, discipline, strength, courage, and authority. Related concepts include virility and valor, along with warrior bravery and respect for lawful authority. The Roman religion ensured that male gods exemplify these virtues, and political and military leaders idealized them. Roman intellectuals, moderated by Greek philosophy, adjusted the list of virtues towards features of wisdom, such as prudence and justice. The Roman statesman Cicero (106–43 BCE) says in *De Inventione* (II, LIII): "Virtue may be defined as a habit of mind (animi) in harmony with reason and the order of nature. It has four parts: wisdom (prudentiam), justice, courage, temperance."

Early Christian Fathers adopted the Roman virtues. St. Ambrose (330s–397 CE) named the cardinal virtues of temperance, justice, prudence, and fortitude, and these four were supplemented by the three spiritual virtues of faith, hope, and charity. Jesus has sometimes been depicted over the centuries as more "feminine," to emphasize submission, selfless service, and self-sacrifice. Yet Jesus can just as well be depicted as God's warrior, battling against sin on earth and against Satan /Anti-Christ when Armageddon comes.

Judaism and Christianity are hardly the only religions who expect their gods to do holy battle. The Persian religion of Zoroastrianism, so influential on later Middle Eastern religions, depicted the entire history of creation as one long battle between the good god Ahura Mazda and the evil god Angra Mainyu. Like the Greeks, and Nordic tribes, the early Hindus depicted their pantheon of gods (perhaps all descended from an earlier pantheon of Aryan deities) as sharply masculine or feminine. The male gods—such as Varuna, Agni, and Indra—are distinctively built for power, command, and war. Idealized demigods of Chinese mythology are typically kings and warriors from long-ago "golden" eras. Aztec and Mayan male gods are always ready for war.

Virtues encouraging a militaristic spirit are rarely absent from religion. Warlike conduct between peoples has only rarely been regarded by religions as deviant or sinful behavior. The rare religious reformer or saint can insist that the Creator made us to be peaceful, but religions don't listen much. Religious scripture, religious practice, and religious law almost always approves of war because the gods approve of war and enlist their religious followers in holy wars. Only some militaristic and fascist religions assume that war is naturally good for humanity, but most religions are able to regard the human tendency to go to war as normal and natural, and sometimes regrettable too.

Religions are not innately pacifist. Every significant religion (setting aside tiny churches and cults) that has been heard preaching pacifism has also been a religion that has gone to war when necessary. Major religions from the Egyptian, Persian, and Roman cultures to the Indian and Chinese civilizations have all endorsed war as necessary for fulfilling divine designs. Christianity and Islam read how their scriptures are full of holy war led by their gods, and both developed traditions of "just war" theory which explain how to justify going to war when times demand aggressively defensive action.

Why can't there by "virtue" in staying home...?

Because the world's major religions all arose after agriculture and animal domestication, most of them easily justify humanity's "war" on other species. Prominent exceptions come from India: Hinduism and Jainism, along with much of Buddhism. Ancient Judaism didn't appear to practice vegetarianism. The Bible approves of meat eating in many places. Curiously, the first Genesis passage about God's instructions to the first humans suggest vegetarianism. Genesis 1:29–31: "And God said, 'Behold, I have given you every herb bearing seed which is upon the face of all the earth, and every tree in which is the fruit of a tree yielding seed; to you it shall be for meat.'" But Genesis 9:3 says, "Every moving thing that liveth shall be meat for you."

The human war against the environment has not been regarded by most religions as deviantly unnatural and sinful behavior, either. Scriptures either give dominion over the earth to humans, or have nothing to say about large-scale effects of human activity on the planet's resources and forms of life. Religions look backwards to inspired origins by gods, or far forwards to the end of the world, but hardly ever have anything to say about good management of the planet now. Even where a religion encourages the wise management of local resources, so a tribe doesn't take more than its natural share, few other religions care to listen. The world's major religions have not prioritized slowing down humanity's race to exhaust nonrenewable resources or humanity's devastating eradication of the world's habitats and species. We'd be able to tell when major religions get serious; they have the means to preach that heaven is unattainable to those who help pollute and destroy our precious nature. Yet almost all religions are still focused on telling people what must be done to escape from this earthly world.

Notes

1. McCleary, D. F., C. C. Quillivan, L. N. Foster, and R. L. Williams, "Meta-Analysis of Correlational Relationships Between Perspectives of Truth in Religion and Major Psychological Constructs," *Psychology of Religion and Spirituality* 3.3 (2011): 163–180.

2. Green, Martha, Don S. Browning, and Christian John Witte, *Sex, Marriage, and Family in World Religions* (New York: Columbia University Press, 2013), pp. 242–243.

3. Ibid., p. 29.

4. Abdullah Yusuf Ali, trans., *The Holy Qur'an* (1934 edn). Compare with the version by M. A. Abdel Haleem, trans., *The Qur'an* (Oxford: Oxford University Press, 2004), p. 54.

5. Edward J. Latessa and Paula Smith, *Corrections in the Community*, 5th edn. (Burlington, Mass.: Elsevier, 2011), chap. 2.

6. Bernard de Mandeville, *The Fable of the Bees* (Edinburgh: printed for J. Wood, 1772), p. 216.

Further Reading

Bellah, Robert N. *Religion in Human Evolution: From the Paleolithic to the Axial Age.* Cambridge, Mass.: Harvard University Press, 2011.

Boyer, Pascal. *Religion Explained: The Evolutionary Origins of Religious Thought.* New York: Basic Books, 2001.

Dennett, Daniel. *Breaking the Spell: Religion as a Natural Phenomenon.* New York: Penguin, 2006.

Goldberg, Ann. *Sex, Religion, and the Making of Modern Madness.* Oxford: Oxford University Press, 1999.

Greenawalt, Kent. *Does God Belong in Public Schools?* Princeton, N.J.: Princeton University Press, 2005.

Hawley, John S. *Sati, the Blessing and the Curse: The Burning of Wives in India.* Oxford: Oxford University Press, 1994.

Hinshaw, Stephen P. *The Mark of Shame: Stigma of Mental Illness and an Agenda for Change.* Oxford: Oxford University Press, 2006.

Landau, Ruth, and Eric Blyth. *Faith and Fertility: Attitudes Towards Reproductive Practices in Different Religions from Ancient to Modern Times.* London and Philadelphia: Jessica Kingsley Publishers, 2009.

Leslie, Julia. *Roles and Rituals for Hindu Women*. Delhi, India: Motilal Banarsidass, 1992.

Popkewitz, Thomas S., ed. *Rethinking the History of Education: Transnational Perspectives on Its Questions, Methods, and Knowledge*. London: Palgrave Macmillan, 2013.

Reagan, Timothy G. *Non-Western Educational Traditions: Alternative Approaches to Educational Thought and Practice*, 3rd edn. Lanham, Md.: Lawrence Erlbaum, 2005.

Richards, David, and Nicholas Bamforth. *Patriarchal Religion, Sexuality, and Gender*. Cambridge, UK: Cambridge University Press, 2008.

Schumaker, John F., ed. *Religion and Mental Health*. Oxford: Oxford University Press, 1992.

Selengut, Charles. *Sacred Fury: Understanding Religious Violence*. Lanham, Md.: Rowman & Littlefield, 2008.

Siker, Jeffrey S., ed. *Homosexuality and Religion: An Encyclopedia*. Westport, Conn.: Greenwood Press, 2007.

Stamos, David N. *Evolution and the Big Questions: Sex, Race, Religion, and Other Matters*, 5th edn. Malden, Mass.: Blackwell, 2008.

Welch, Michael R., David Sikkink, and Matthew Loveland. "The Radius of Trust: Religion, Social Embeddedness and Trust in Strangers." *Social Forces* 86 (2007): 23–46.

Wright, Robert. *The Evolution of God*. New York: Little, Brown, and Co., 2009.

6

GOOD AND EVIL

Religions all have definite views about right and wrong. They try to grapple with the potential depravity inside of us and the many evils that surround us. Religions offer ideas about the source of human morality, and the basis of goodness, evil, aggression, love, empathy, righteousness, and justice. Are there demonic powers, and could there be some sort of cosmic justice? Why do bad things happen to good people? Science's evolutionary standpoint on the ability of humans to be moral and to work for the common good has added needed clarity to these questions.

Evil

People need little help recognizing evil when they experience it themselves, or see it happen to others. Evils are those terrible events that cause suffering, destruction, and death to innocent people, events for which little or no good could come from them, so that we can only view them with dread and regret. No religion, and no science, is needed to figure out obvious cases of evil; people only need their common sense and sensitive hearts. You surely know evil when you see it.

There is nothing more common among the world's religions, past and present, than embodiments of evil. Malevolent gods, demons, monsters, and ill-tempered spirits populate the mythologies of ancient religions that originated from the first civilizations. Earlier tribal religions predating the first civilizations didn't leave behind their stories in recorded form, since they lacked writing. But it is difficult to suppose that the first evil gods had to await the invention of writing in order to awaken in the human imagination. The earliest religions among hunter-gatherer tribes were probably replete with tales about dangerous deities as well as helpful deities. Perhaps religion was born from fear as much as from hope. Surviving indigenous peoples around the world have very old myths about the earliest times of creation, stories that require multiple gods to explain the dramatic action, and those gods aren't exactly the purest moral beings. Instead, they are highly anthropomorphic deities who can exemplify the worst in human character as well as the best. The first religions of the early civilizations that left written records about their gods similarly describe groups of gods who weren't expected to be entirely beneficent.

None of these religions could be accused of overlooking evil or trying to explain away evil. Evil was always pretty much front and center. Explanations for terrible disasters striking entire populations, or horrible tragedies affecting individual people, were never hard to come by. It could be fairly judged that religions were in the business of explaining evil, and explanations weren't difficult to supply, since divine powers of the world could easily turn hostile and inflict suffering on humanity. Religions profited from explaining evil in countless ways. Religions have always been kept very busy offering predictions of evils that could happen, supplying formulas for preventing potential evils from happening, explaining evils when they do happen, and suggesting remedies for evils for people to undertake in order to sooth anxiety and relieve grief. There's little mystery about evil in all these religions. Evil is built into the fundamental nature of the world, and into the essential nature of many of the spirits infesting the earth and the gods inhabiting the heavens. There was little incentive for religions to overlook or explain away evil, as if religious believers shouldn't be focused on the evils that visit everyone eventually. Evil has kept religion in business.

Many of the world's indigenous peoples retain beliefs about spiritual beings active here among us, and they credit certain people, labeled as "shamans" by anthropologists, who can understand and interact with those spirits. The priests of modern religions aren't that different from shamans. They all have special knowledge about unnatural affairs and agents; they can detect not only the activities of such unnatural things but also predict what they might do; they know the spiritual vulnerabilities of human beings to these unnatural powers; and they wield magic to deter those unnatural forces from doing harm or at least help repair harms done. The tribal shaman uttering incantations to strengthen a sick person's derelict spirit in those hopes of warding off demonic harms isn't different in basic form from a Catholic priest guiding a sinner's remorse towards the soul's regeneration to prevent a god from inflicting a torturous afterlife.

In every religion, the same message resounds: there are grave evils lurking and awaiting our weaknesses for exploitation—beware!—and your weakness can be aided by priestly interventions with those evils. There is no end of evils in the world, each religion explains, and many of those evils are your own fault, for you to deal with, and for you to manage with religious assistance.

Divine Evil

Indigenous religions having long histories going back into prehistoric eras are never lacking in all sorts of spirits and deities requiring close attention. The first civilizations had polytheistic religions about vaguely anthropomorphic or animallike gods, any one of them capable of helping or harming humanity as their divine judgment dictated. In the West, religions of civilizations from Iran to the Mediterranean described gods willing to deal with humans on their divine terms, so the people had to show proper obedience in order to receive divine favor and goodness. Dealing with a Yahweh, for example, was a most hazardous relationship, since it was obvious to the Hebrews that no end of evils could be delivered if a covenant was violated. In the East, religions from India to Japan came to view the various gods as manifestations of an even more fundamental divine reality, but that reality contained all things and qualities, and couldn't be said to be entirely good or evil. The Hindu god Brahman and the Chinese "god" called the Dao weren't categorized as mostly beneficent, or malevolent, but somehow impersonally beyond good and evil.

Both Hinduism and Buddhism maintain vast compendiums classifying all sorts of angelic and demonic spirits that populate this world and other worlds. Chinese folk religion, which significantly overlaps with popular Daoism, isn't far behind in the enumeration of spiritual forces such as ancestral deities, nature gods, and demonic spirits. Christianity's reverence towards thousands of demigods wielding unnatural powers (called "angels" and "saints") is a diminutive version of this same religious feature, and Christianity was never behind any other religion in its recognition of vast zoos of devils and demons.

Sorting out supreme gods who were entirely on the side of good, or of evil, was a theological development that occurred later, during the Axial Age of roughly 800–200 BCE. The first theological religion to attempt to isolate all that is truly good within a single god was the Persian religion of Zoroastrianism. Its ancient roots in the sacred hymns of the *Gathas* speak of two opposed cosmic deities, one for good and one for evil. By 600 BCE, Zoroastrianism was consolidated and codified in a larger holy book called the *Avesta*, which declares that these two gods are the only gods, they had always existed, and they have been in cosmic conflict since eternity. The perfectly righteous god, Ahura Mazda, cannot be faulted for any evils. However, since no religion can ignore

Ahura Mazda and Angra Mainyu do the dishes

evil, Ahura Mazda had divine competition, in the form of a second god, Angra Mainyu. These two gods are the first in human imagination to have a co-original status (neither one was created) with completely opposite moral characters: one representing pure goodness and justice, and the other representing pure evil and chaos.

Not even the "darkness" and "light" of Daoism's Yin and Yang have that rare theological status, since Daoism holds that both Yin and Yang contain mixtures of the other within itself, and both qualities originate from a more fundamental reality, the Dao. Furthermore, unlike Daoism's recommendation that human welfare depends on maintaining a harmonious balance of Yin and Yang in one's life, Zoroastrianism demands that each person must decide which god to follow, the good god or the evil god. This choice comes with a dire warning: the followers of the evil god are destined to lose in the coming cosmic battle for the fate of the universe, because Ahura Mazda and the forces of good will eventually be completely victorious.

Another Axial Age religion, the priestly religion of monotheistic Judaism, also regarded one primeval god, Yahweh, as essentially good. Judaism's earlier phases of polytheism, when the Hebrews worshipped several gods, was followed by a period of henotheism when only Yahweh was worshipped instead of other available gods. Those kinds of religions were replaced with monotheism—faith that only one god is real—by 500 BCE, as the impact of the Babylonian Captivity (590s–510s BCE) was absorbed. Intellectual influences arriving from the Babylonian civilization predated the captivity, but immersion within a Zoroastrian culture made a permanent impact on the educated leadership of the Jews surviving the captivity, who returned to Jerusalem to rebuild everything including their holy books. As *The Jewish Encyclopedia* recounts in its article about Zoroastrianism, the fundamental similarities between the two religions cannot be coincidences. Some of these common features predate the Babylonian captivity, while others are probably due to the captivity itself, as Zoroastrianism had evolved by then as well into a strict dual-god theology:

The points of resemblance between Zoroastrianism and Judaism, and hence also between the former and Christianity, are many and striking. Ahuramazda, the supreme lord of Iran, omniscient, omnipresent, and eternal, endowed with creative power, which he exercises especially through the medium of his Spenta Mainyu ("Holy Spirit"), and governing the universe through the instrumentality of angels and archangels, presents the nearest parallel to Yhwh that is found in antiquity. But Ormuzd's power is hampered by his adversary, Ahriman, whose dominion, however, like Satan's, shall be destroyed at the end of the world. Zoroastrianism and Judaism present a number of resemblances to each other in their general systems of angelology and demonology, points of similarity which have been especially emphasized by the Jewish rabbinical scholars Schorr and Kohut and the Christian theologian Stave. There are striking parallels between the two faiths and Christianity in their eschatological teachings—the doctrines of a regenerate world, a perfect kingdom, the coming of a Messiah, the resurrection of the dead, and the life everlasting. Both Zoroastrianism and Judaism are revealed religions: in the one Ahuramazda imparts his revelation and pronounces his commandments to Zarathustra on "the Mountain of the Two Holy Communing Ones"; in the other Yhwh holds a similar communion with Moses on Sinai. The Magian laws of purification, moreover, more particularly those practised to remove pollution incurred through contact with dead or unclean matter, are given in the Avestan Vendïdād quite as elaborately as in the Levitical code, with which the Zoroastrian book has been compared (see Avesta). The two religions agree in certain respects with regard to their cosmological ideas. The six days of Creation in Genesis find a parallel in the six periods of Creation described in the Zoroastrian scriptures. Mankind, according to each religion, is descended from a single couple, and Mashya (man) and Mashyana are the Iranian Adam (man) and Eve. In the Bible a deluge destroys all people except a single righteous individual and his family; in the Avesta a winter depopulates the earth except in the Vara ("enclosure") of the blessed Yima. In each case the earth is peopled anew with the best two of every kind, and is afterward divided into three realms. The three sons of Yima's successor Thraetaona, named Erij (Avesta, "Airya"), Selm (Avesta, "Sairima"), and Tur (Avesta, "Tura"), are the inheritors in the Persian account; Shem, Ham, and Japheth, in the Semitic story. Likenesses in minor matters, in certain details of ceremony and ritual, ideas of uncleanness, and the like, are to be noted, as well as parallels between Zoroaster and Moses as sacred lawgivers.[1]

Christianity retained nearly all of these shared religious beliefs, including the popular belief in minor spirits (angels and demons), along with the theological idea that a supremely evil deity, Satan, has immense influence in the world.

Satan's evil manifests itself in the human world largely through deception and temptation, as Satan recruits followers who stray from obedience to God and live sinful lives. Zoroastrianism continued to influence Christianity and other religions in the Near East for centuries. A new Iranian prophet, Mani (c.216–276) combined Zoroastrianism with Christianity to create Manichaeism, which continued to insist that twin gods, one good and one evil but neither having omnipotent powers, are locked in a struggle for the fate of the cosmos. After

Catholicism emerged supreme by the sixth century CE in the West, it continued to label as "Manichean" any worldview that saw only pure goodness and pure evil at work in the world. Islam also uses the idea of Satan, who is called Iblis or Shaitan. Supreme among the treacherous spirits called jinn, which Islam didn't invent but inherited from earlier tribal religions in the Arabic region, this Saytan rebelled against Allah and attempts to lure both lesser jinn and humans into sin. The Qur'an (sura 114: 1–6, trans. Yusuf Ali) describes the problem presented by this Satanic "whisperer":

> Say: I seek refuge with the Lord and Cherisher of Mankind,
> The King (or Ruler) of Mankind,
> The God (or judge) of Mankind,
> From the mischief of the Whisperer (of Evil), who withdraws (after his whisper),
> (The same) who whispers into the hearts of Mankind,
> Among Jinns and among men.

Even as stringent a monotheism as Islam makes room for evil forces that aren't under the good god's control. Polytheisms that recognize an evil god, or a monotheism that doesn't insist on god's complete goodness or supreme power, can explain how evil persists within creation. But why does a completely good and all-powerful god tolerate evil forces? Monotheistic religions have tried every conceivable way to account for this odd situation. But the terms of the problem cannot be changed: evil is quite real, evil is caused by evil powers, and those evil powers are somehow subordinate to the lone supreme deity. Somehow the supremely good god needs evil powers to do things that it cannot do itself. Any monotheism, such as Enlightenment deisms or modern Christian theism, which clings to viewing creation as "the best of all possible worlds" that a perfect God could have created, is a monotheism that practically declares evil to be illusory. If you think you see evil in the world, you are simply ignorant, for you do not understand (unlike the theologians) how God regards that apparent evil as a sufficiently good portion of a supremely perfect design for all creation. Lay believers have never found this theological stance acceptable, instead preferring to take their perception of evil as accurate and viewing their god as a warrior god against the quite real forces of evil. The way that typical Christians stubbornly insist on a Manichean picture of evil's genuine powers is ample evidence that the practical mind won't be told that the problem of evil can be erased by erasing evil. Evil must be real—that's the very point of needing a good and powerful god on your side.

Every religion is fundamentally about evil as much as it is about unnatural deities and demons that humanity must deal with. What would happen to the religion, hypothetically, which had a message saying that there are no serious evils, only good things in this life, and in regard to the annoying sufferings people must endure, there is nothing practically to be done about them and you are on your own. If there ever was such a religion, it expired and vanished for lack of followers. People are practical minded, after all. People want change, they want to make a difference, they want help making their lives better, and they want relief from suffering and despair. And they will follow whatever will supply their wants. It is no failure of practical intelligence to seek relief from earthly troubles and griefs. If religion is to be faulted for any intellectual failures, those failures must be about religion's fictional means of providing aid to humanity, and exaggerating the amount of aid that humanity really requires.

Human Evil

How much assistance does humanity actually require? There is nothing more familiar to the priestly recipe for relieving the human condition than to place blame on the human victim as much as on evil spirits. Every religion will tell you that evils are a basic aspect of the earthly world of troubles and woes. Religions are just as likely to tell you that there are evils inside of you, too, that make you more susceptible to these hazards.

Every religion on earth has utilized one or more of these simple recipes: (1) You committed some evil act, worthy of unnatural punishment; (2) You failed to observe due respect or subservience to some unnatural being, bringing its disapproval upon your head; (3) You failed to contribute to your group's favorable status with some god, so that god has withdrawn its protection (either from you, or perhaps from the whole group); or (4) You can't fulfill expected standards of thought or deed, rendering you vulnerable to suffering in this world and probably the next world as well.

Immorality, impiety, infidelity, and imperfection are innate human faults rendering us not only susceptible to, but deserving of, unnatural harms from spiritual powers. The more deities a religion has, the less that religion's followers have to worry about any particular god, and if those gods are themselves imperfect, then perfect behavior is hardly expected of ordinary folks. Indigenous religions dating from prehistoric times threaten harms in this life for ignoring the gods, but conformity and obedience is achievable with reasonable effort. If the supreme spiritual reality isn't even like a mind or a person, then it can't personally care about what you do in your life, and any suffering you endure for living a poor life is pretty much self-inflicted. The religions of China, Korea, and Japan don't regard perfection as necessary at all. India's Hinduism, and Buddhism across Asia both predict many reincarnated lives for even the best of people before anything like spiritual perfection could be attained. Only the monotheistic religions, notably Christianity and Islam, hold people to extremely high standards and simultaneously threaten eternal suffering in an afterlife for failures in this one earthly life.

Why does this correlation between monotheism and perfectionism exist? Theologically, it must have something to do with the problem of evil. In both Christianity and Islam, the supreme God is morally perfect, but human beings are truly special creatures too, since they possess the spiritual capacity to freely choose good or evil, which means that they possess the divine ability to create goodness or create evil that the supreme God does not create. Yahweh's original creation of the perfect Garden of Eden was destroyed by humans, who radically altered Nature forever by their choices. This means that the imperfect Nature we now live in was created more by humans than by God, so that it is theologically correct to regard humanity as cocreators of nature. Because humans are entirely responsible for the evil world we now inhabit, God is relieved from responsibility for evil in the world. The first humans could have resisted the lure of tempting Satanic lies, but they didn't, so the resulting natural sufferings that humanity must endure are entirely their own fault.

Seriously... ?

Blaming humans entirely for creating evil generates the concept of sin. Sin is possible only within a religious framework that takes the one God to be perfectly righteous so that deeds by humans deviating from that divine righteousness can only be terribly evil, and hence deserving of the worst punishment possible. A very bad person of a different religion, a religion about a lone God (or a group of gods) that lacks moral perfection, could only be a very bad person, not a sinner. How could a bad person be judged so harshly by those who can do things just as bad? But a morally perfect God setting the required standards can be the harshest judge. A comparison with Hinduism and Buddhism is again helpful, because those religions do judge bad people quite severely, but they don't use the concept of sin to express moral judgment. Instead, they blame bad people for creating negative karma within their souls, causing rebirth into worse worlds. However, Hinduism and Buddhism don't regard lower worlds, some of them very much like unpleasant hells, as eternal jails for irredeemable souls. Souls that learn from punishment and accumulate enough positive karma for obedience in these hells are naturally reborn into higher and better worlds eventually. Since there is no supreme deity in Hinduism or Buddhism who sets down moral laws and renders judgments on humanity, only each person's accumulation of positive karma controls how high one's sequence of rebirths can be elevated, potentially all the way up to maximal bliss in unity with Brahman (for Hinduism) or into complete release from desire and suffering into Nirvana (for Buddhism).

The concepts of monotheism and moral perfection are responsible for the notion of sin, and they are also responsible for the problem of evil. The question is easy to ask: Why does evil exist when a perfect God could prevent it? The foursome of God, Satan, Humanity, and Nature are together capable of nearly resolving the problem of evil. God is both perfect and a creator; Satan creates nothing but can tempt a free will; humanity is a creator and has free will; and nature is created and imperfect. Evil is real for three reasons: nature is imperfect (which we easily see), humanity can do evil (obvious as well), and human free will can be swayed by evil. Unless there was some sort of evil besides humanity from the start, there'd be no good explanation why free will would think of evil and choose evil, and humanity would easily find God's goodness perfectly sensible (so humanity would end up just like angels and never leave Eden). This explains why Satan is needed in this story, but nevertheless humanity remains responsible for its own evil and for nature's evils (and hence deserves punishment). Human evils and natural evils are thereby accounted for. Ultimately, the tension in the problem of evil comes down to the relationship between God and Satan. If the dualistic Manichean solution is used, then God remains perfect but isn't all-powerful, so it takes great effort by God to eventually defeat Satan. However, in a monotheistic religion, the question persists: If God is all-powerful and hence could have easily prevented Satan from doing harm, why does God let Satan prevail for so long?

Here, the connection between monotheism and perfectionism is entirely exposed: unless Satan is permitted to tempt humanity as much as God, humanity would find it very easy to be morally perfect, since the attractiveness of God and God's rewards would always sway free choice in the absence of any serious competition. However, humanity cannot be morally perfect, yet we ought to be. We obviously don't look morally perfect, and we don't think that everyone deserves God's approval. But Islam and Christianity hold people to much higher standards: a single grave sin, committed just once, is sufficient to warrant Hell forever. If humanity weren't expected by God to be perfect, and most people could easily satisfy moderate expectations, then God would be satisfied with the prevalence of evil within creation. But what sort of God would be satisfied with a creation that included lots of evil? If this God is the supreme creator, as monotheism demands, then satisfaction with lots of evil in creation implies that such a creator God cannot be perfectly good and all-powerful. On the other hand, if there is only one good God who is never responsible for evil, then that God must hold humanity to the highest expectations of goodness, so that evil can both plentifully exist and be entirely created by humanity.

Great theological importance is therefore attached to human free will. God may be responsible, ultimately, for the existence of Satan, but Satan does not have any powers of creation, only the ability to voice temptations to free wills. But because a person's will is free, Satan's deceptive temptations cannot control that free will to do anything—only each person does anything by his or her own free choice. Satan is necessary in this explanation, since unless the free will was tempted, it would usually choose the good. God can therefore be responsible for Satan's existence—since God let Satan exist and could eliminate Satan—but God isn't responsible for evil since God isn't responsible for anything Satan does. Satan has his own evil will and can't create evil anyway—only humans can do that. Because there are (obviously) so many evils, and humans are responsible for all of them, then humanity is evidently being held to an extremely high standard in order to fail so spectacularly. Think of the doctrine of Original Sin in Christianity— all of humanity must be horribly sinful and creators of terrible evils due to ancestral choices, and it was one choice, in fact, to disobey a single instruction from God. Islam lacks the doctrine of Original Sin, but its approach to the problem of evil is quite similar to Christianity's. Islam holds that each person is born innocent, but everyone is quite susceptible to the temptations of evil, and hence people are entirely responsible for the evils they commit.

Neither Christians nor Muslims are in the habit of questioning why humanity is held to perfectionist standards of conduct. Plenty of other religions uphold high moral standards, exemplified by idealized human figures—demigods, saints, prophets, and so forth—but they don't really expect each person to be exactly like those models, much less like God. Why must only sinless people be worthy of acceptance by the supreme Creator? A perfect God must uphold perfect standards somewhere, it must be supposed, lest this God rest content with a less than perfect (partially evil) creation. God figured out the solution to this tough problem of evil. And it is a tough problem, even for a God. How can a perfect God create something that is less perfect, without taking the blame for creating evil? A smart God can apparently figure this one out: create frail limited beings like humanity, cripple their judgment with free will, let the tempting voice of evil sway them into sin, and then blame all evil on them when they fail to be perfect.

Does that actually sound like a reasonable plan?

Human Goodness

Neither science nor religion is needed by anyone in order to judge when someone does something good. If any evidence is needed to justify this common sense fact, there is plenty. For example, there are innumerable people who know nothing of science or of religion who nevertheless learn how to be moral during their childhood and behave fairly morally as adults. The great teacher of morality is culture, not science or religion, and culture is transferred primarily in childhood within the social settings of the family and neighborhood. Since science does not pretend to be the author or transmitter of cultural morality, we can focus on religion here. Religions all claim to be the upholders of sound morality, the teachers of sound morality, and the best judges of immoral violations. When confronted with evidence that nonreligious people have no difficulty acquiring ordinary morality and staying as moral as adults (at least as moral, on average, as religious people), the religious reply is that culture ultimately can be moral because of religion's involvement. So long as the wider culture is indebted to religion, the story goes, then it is possible for some people to become moral from that culture, even though they avoid religion itself.

The relationship between culture and religion must be sorted out in order to see where religion is truly necessary for morality, if it is necessary at all. If religion is necessary for moral culture, as a hypothesis for consideration, then certain things would be true. For example, it would be true that every culture has religion controlling its morality. But that isn't true. There are societies around today with low religiosity—Scandinavia and Eastern Europe, for example—that show no signs of moral collapse. And in the distant past, when religions first arose, they arose within human societies that already had plenty of morality. This is so, for two main reasons. First, religion arose after the invention of language, since mythic narratives require some language, and full language is probably less than 70,000 years old. That sounds like a long time ago, but Homo sapiens had been already living in small groups of less than fifty individuals with intensely cooperative social bonds for one hundred thousand years since the species' evolutionary origin, so they were using moral rules to enforce good behavior among them. Any significant amount of anarchy or chaotic violence within small groups of people, surviving only by being so mutually dependent on each other, couldn't have been practical.

Second, even if the first religions somehow invented the first moralities, what are the odds that all of the world's religions would have invented pretty much the same basic set of moral expectations and virtues, like honesty, fairness, empathy, altruism, and so on? Religions rarely agree about much, telling tales about wildly different gods and creation events. Expecting them to agree on anything they dream up is pretty farfetched, so religion wouldn't have invented morality. The only way to explain why every human society all around the world exhibits about the same common core morality is to locate morality's origins in humanity's distant past, during the species' own evolutionary origins. This inference relies on a basic principle of anthropology: if you find something that almost all humans are doing pretty much the same way, then it evolved a very long time ago when there were few humans, they lived in the same geographical region, and they were more closely related to each other. Humanity had evolved in Africa to use one basic morality, easily taught to children and expected of all adults. When religions were invented later, each religion approved of the morality already there in culture. That is why it is so easy to teach small children everything they need to know about ordinary morality and virtue: their developing minds can easily understand what is expected of them without any other information (like religion) involved at all.

Each religion itself is able to at least suspect that it isn't responsible for morality. After all, not only does every religion teach pretty much the same basic morality, when religions supplement that morality with additional rules for adult society, it soon finds that the religions of neighboring cultures teach very different rules. Why are religions disagreeing over these additional rules? On the other hand, if one's culture has little contact with different cultures, it is very easy to suppose that one's own social rules are the only rules, uniquely authorized by one's own gods.

The experience of ancient Greek contact with different cultures is instructive. Before the Greek peoples learned much about what else was going on around the Mediterranean, their own gods were enough. The Greek poet Hesiod wrote his *Works and Days* around 700 BCE, expressing the common belief that rules about what is good and right for all humanity was dictated by the supreme god, Zeus.

> For the son of Cronos has ordained this law for men, that fishes and beasts and winged fowls should devour one another, for right is not in them; but to mankind he gave right which proves far the best. For whoever knows the right and is ready to speak it, far-seeing Zeus gives him prosperity; but whoever deliberately lies in his witness and foreswears himself, and so hurts Justice and sins beyond repair, that man's generation is left obscure thereafter.[2]

Two and a half centuries later, the Greek historian Herodotus expressed a very different view of human morality. Herodotus was familiar with much of the known world, as many educated and well-traveled Athenians were by that time. After all, Athens had become a merchant capitol city with a cosmopolitan atmosphere, where the entire Mediterranean world and much of the Middle Eastern world came to do business. His *Histories*, composed between the 450s and 430s BCE, recount many detailed episodes that may or may not have happened exactly as Herodotus recorded, but he is always clear about what his stories do explain about human nature. This tale is typical of his attitude towards morality:

> ... if it were proposed to all nations to choose which seemed best of all customs, each, after examination, would place its own first; so well is each convinced that its own are by far the best. It is not therefore to be supposed that anyone, except a madman, would turn such things to ridicule. I will give this one proof among many from which it may be inferred that all men hold this belief about their customs. When Darius was king, he summoned the Greeks who were with him and asked them for what price they would eat their fathers' dead bodies. They answered that there was no price for which they would do it. Then Darius summoned those Indians who are called Callatiae, who eat their parents, and asked them (the Greeks being present and understanding through interpreters what was said) what would make them willing to burn their fathers at death. The Indians cried aloud, that he should not speak of so horrid an act. So firmly

rooted are these beliefs; and it is, I think, rightly said in Pindar's poem that custom is lord of all.[3]

Any historian of world societies can similarly see how culture is the genuine power over social custom, not religion, while each religion tries to preserve its culture's traditions. The power displayed by culture is nevertheless more impressive in the long run.

When a culture tries to progressively modify and improve its social rules, it usually encounters resistance from religious forces doing what they are designed to do: prevent deviations from traditional ways and morals. If the forces of cultural progress are strong enough, some progressive religious people may tear away and divide the religion over the issue. The way that cultures can undergo moral progress—for example by freeing slaves, empowering minorities and women, and protecting children from exploitation—without uniform encouragement from established religion, proves beyond doubt that religion isn't responsible for morality. If religions were always responsible for morality, then we would see entire religions at the progressive forefront of all social progress, instead of reluctantly following progress from behind, as it happens. (Although religions make sure to take credit for that progress long afterwards.) Baptists, Presbyterians, and Methodists in twenty-first century America, for example, hardly realize today how the idea of freeing the slaves bitterly divided their denominations in the nineteenth century.

It is also easy for today's religion to see how its social rules for the adult world aren't respected in other cultures. The shock of encountering a different culture, with a different religion that strongly defends quite different cultural folkways, arouses confused and hostile reactions. What is a religion to think? Ultimately, there are four options for a religion to choose from: (1) universalism—our religion is entirely correct, so any other culture is morally wrong, and has the wrong religion; (2) religious relativism—our religion is entirely correct for our culture, and our god(s) only require our fidelity, not the fidelity of foreigners; (3) cultural progressivism—our culture presently follows a worthy morality, letting us know what is best in our religion and what can be ignored; and (4) subjectivism—religious individuals may keep supposing that their god(s) are responsible for morality, but culture is actually entirely responsible for morality. The first option is preferred by dogmatic and fundamentalist followers of monotheistic religions such as Christianity and Islam. The second option is practically adopted by cultural religions such as traditional Judaism, Hinduism, Buddhism, and Daoism. These religions aren't fixated on converting other peoples and changing other cultures; if other people find worthy ideas and practices in this second kind of religion (such as meditation), they are welcome to adopt and adapt them. The third option is taken by a modernized religion such as (most of) Christianity in Europe and America, and (some of) Hinduism in India. Modernized people know right and wrong when they read it in their holy scriptures, and that's why they don't have to think that scriptures are infallible about morality, and therefore they don't suppose that God really requires them to follow every rule they read in the Bible. The fourth option is selected by Secularism: the view that religion is best left as a private matter and it shouldn't be permitted to control social morality or politics.

Modern religious people mostly fall into the culturally progressive category, since they aren't comfortable with damning all other peoples of the world, as universalists do, and they don't assume that everything about their religion has been forever correct. Confronting today's Christians with terrible evils in the Bible done by Yahweh's "Chosen People" such as mass genocide, murdering innocent people, and horrible abuse of children, women, homosexuals, and slaves, all approved by Yahweh in the Old Testament, doesn't seem to trouble many of today's faithful. They may simply assume that later Christianity exemplifies a higher morality, exemplified by Jesus's example in the New Testament. Confronting Christians again, with New Testament endorsements of things like the subordination of women to men, prejudice against homosexuality, and slavery, typically makes a Christian retreat to just the sayings of Jesus alone, abandoning most "morality" in the Bible along with much of the gruesome history of the Church. At this stage, cultural progressivism is obviously at work, since such Christians are using what they believe is currently right about their culture today in order to judge for themselves the morality of what they read in the Bible.

When it comes to judging human goodness and explaining why people can know right from wrong, religion has little use. Religions impose their own doctrinal images of human nature on people just to help reinforce the urgency of depending on religion. And when religions happen to hit upon appropriate standards of right and wrong, it is either because they are echoing the basic morality the entire species utilizes thanks to evolution, or they have upgraded their standards to follow more progressive thinking going on in their cultures.

Natural Goodness

If religion is neither the source of morality, nor a reliable judge of what is morally best, does that make science a better authority?

Reading science to glean rules of moral conduct can't work at all, so science never pretended to be delivering moral commandments. Philosophical-minded admirers of science have tried to deduce moral rules for humanity from the natural laws of the world. Those deductions have proved to be disastrous, lending credence to grossly unethical principles behind nineteenth-century ideas like Social Darwinism (let the unfit poor die) and Fascism (only the strongest should survive). Human beings aren't what we thought they were, though. Outdated notions that a human being must be very selfish at the core haven't held up in the light of current biological, anthropological, and psychological research. The only place where loud warnings about the egotistical and cruel nature of every human being is still frequently heard is theology, not any scientific field. Why would some theologians still be interested in rejecting science in order to claim that, if they were left on their own, people would be evil monsters towards each other? The previous section has already exposed the many incentives that a religion dealing with the problems of evil has to place all the blame on humanity. Besides, if humanity could be fairly good all on its own, what would happen to religion?

It is entirely unnecessary to contradict biased theology by assuming that humans must instead be naturally good. The nineteenth century also witnessed philosophical rebellions against religion that romanticized human nature into almost angelic proportions. Many utopian communities were founded on the idea that if only people escaped from stifling negative religion and distracting city life, they could live in peaceful harmony without government. The brief existence of most of those utopian communities tells us a lot about the practicality of their idealistic hopes. We aren't naturally devils, but we aren't so good, either.

Contemporary atheist enthusiasts of natural science, such as Sam Harris, are still heard proclaiming that science is ideal for determining what would bring more human happiness in the future, or at least that science can figure out how to efficiently deliver what we happen to want now. Just a little reflection can tell us things that careful philosophers have long understood: no amount of scientific facts could add up to discovering whether prioritizing happiness is truly right, and science couldn't explain how to fairly distribute all the material goods from technology. Handing morality upgrades over to science only manages to enthrone bad ethics having nothing to do with science. We should never abandon our responsibility to choose among possible futures, and learn from the resulting consequences for ourselves. Science grasps how we can undertake such crucial enterprises, since evolution made Homo sapiens into an intelligently social animal, and then into a cultural animal making its own rules about structuring societies. The diversity of cultures results in some relativism about morality, but it is foolish to run away from troublesome relativism by running into the arms of science's unethical tyranny.

The scientific study of ordinary human morality by the behavioral and brain sciences has revealed pretty much what we already understood about each other. Morality is contextual, local, and reinforceable. Context matters: a person can be cooperative, helpful, and peaceful when personal and material needs are met and social conditions aren't stressful, so treacherous competition and physical violence make little sense. Locality matters: people are generally far more willing to be generous, cooperative, and fair with those who are in proximity, who are familiar, or those who at least share similarities. Reinforceability matters: people are more moral after training and reinforcement conditioning, and they tend to stay quite moral when there continue to be social consequences to their behavior. When does morality diminish? People are far less likely to do the moral thing when they are stressed by lack of basic material goods and lack of security, or when they feel distant or dissimilar from another person, or when they aren't observed by others in their social groups (and especially when they aren't noticeable at all).

These features of ordinary morality are entirely consistent with one theory about morality: the evolutionary theory that the human capacity for morality evolved along with Homo sapiens's emergence as a highly social animal whose survival depends on much cooperation with everyone in a small clan-sized group. To this day, any young child quickly acquires the core moral habits and virtues from older people who value cooperation, adults continually emphasize moral conduct towards people that a child encounters, and everyone urges conformity to the ways that the social group behaves properly. Ordinary morality therefore suffers from characteristic limitations, as well. When high stress diminishes the value of cooperation, morality fades within a group; morality depends on reinforcing familiarity, and therefore stereotyping as well, and moral motives fade when others in your groups can't see what you're doing. Morality can be lectured to children in the form of universal principles—"Don't hit anyone!"—but morality in practice is taught by example, so that one feels the pang of regret when hitting an acquaintance, but not when one hits a stranger.

Evidently the practical limitations of ordinary morality, strong enough to hold together small human groups, couldn't prevail in large societies. And we don't see it prevailing enough today. Social order in groups larger than tribes, such as kingdoms, nations, or empires, is maintained by government and law as much as morality. Where religion is not restrained by law, and not concerned enough for morality, religion is capable of any barbaric cruelty imaginable. If people think there is a god, nothing is forbidden. The point is not that any religion necessarily will commit horrible immoralities, but rather that any horrible immorality is possible where religion is law. Every significant religion that has ever acquired great power has a list of atrocities recorded next to its name. A religion's adherents are not proud about their church's dark moments in the past, but they think it is easy to defend their religion by blaming atrocities done by their church upon those past barbaric times, not on the religion itself. Yet their church's complicity with those times is precisely what condemns that religion. How moral is a religion when it quietly condones, or even participates in, those barbarities?

Any religion can serve as an example; let us look at one supplied by a prominent Anglican priest of England, William Ralph Inge (1860–1954). Not just any priest, but Professor of Divinity at Cambridge University and Dean of St. Paul's Cathedral. His religion has had a past as dark as any. Its highest leadership, including Popes and theologians such as St. Aquinas, authorized the Inquisition during medieval times. His book *Christian Ethics and Modern Problems* honestly recounts the "ethics" of such godly men:

> The use of torture was expressly ordered by Papal Bulls of 1252 and 1259. By an exquisite hypocrisy, the Church "handed over to the secular arm" ("*ecclesia abhorret a sanguine*") those who were to be burnt; princes who refuse to burn the victims were to be excommunicated. Those who were not tortured but only imprisoned hardly fared better. The Consuls of Carcassonne, in their official protest of 1286, described the prisons of the Inquisition there: "Some are so dark and airless that the inmates cannot tell night from day, and thus they are in perpetual lack of air and in complete darkness. In others are poor wretches in manacles of iron or wood, unable to move, sitting in their own filth, and unable to lie except upon their backs on the cold earth; and they are kept for a long time in these torments day and night. In other dungeons there is lack not only of light and air but also of food, except the bread and water of affliction, which is given most scantily."
>
> Even the Bolsheviks have not quite equalled the reign of terror which lay heavily over Western Europe for centuries, cowing all except the most heroic, and effectually preventing intellectual progress, which cannot exist without freedom of thought and speech. It was a cruel age, and it was almost universally believed that God is

more cruel than the most ferocious earthly tyrant. Of course this belief is itself a proof that cruelty at that time aroused very little moral reprobation; but it gave an effective answer to any who were shocked by the torture and judicial murder of inoffensive citizens, whose lives, as even their persecutors were constrained to admit, contrasted very favourably with those of the most orthodox. Coulton says: "About 1233 a suspect, brought before the tribunal, protested as follows in order to clear himself from all suspicion of heresy: 'Hear me, my lords! I am no heretic. I have a wife, and cohabit with her, and have children; and I eat flesh and lie and swear and am a faithful Christian." No wonder that the last words have been omitted by two Catholic historians of the Inquisition! For the ghastly torments of hell, it was supposed, were decreed by God against all heretics. And so, as Thomas Aquinas argues, "If false coiners and other felons are justly condemned to death without delay by worldly princes, much more may heretics, as soon as they are convicted, be justly not only excommunicated but slain out of hand."[4]

As Dean Inge relates, it was a cruel age, and made even crueler by a Church wielding unholy powers.

Liberating Ethics
There is nothing intrinsic to religion itself that forbids or even abates its capacity for cruelty and immorality, if it has unchecked power. Great secular powers like tyrannical monarchs or totalitarian oligarchies haven't been outmatched by theocratic religions, to be sure. Yet religions are supposed to be more ethical, aren't they? If there were anything intrinsic to religion that could restrain it, at least one shining example would be recorded in world history.

But there is no such example written there. Acquiring great power, every religion has gone on to commit terribly immoral things on as grand a scale as it could get away with. And every religion has declared that it has divine approval along the way. Where in the pages of history is that singular religion recorded, which upon acquiring immense worldly power, has ensured its high god sternly disapproves using that power to oppress infidels, unbelievers, and enemies? Nowhere. Among weak religions, things are very different. There are a few gods, small gods of small tribes and weak nations, who have stood for compassion and justice and good will towards all. Some of those small gods eventually became great Gods of Empires, and their inflation of power was always matched by a deflation of morality and an abundance of cruel oppression.

In Europe, genuine ethics was liberated from religious bondage by secular social forces. The only thing that at least slowed the blood-thirsty God of Christianity by the seventeenth century was the force of law backed by powerful yet tolerant monarchs who decided that enough Christian blood had been shed over who gets into heaven upon death. In 1598, King Henry IV of France issued the Edict of Nantes, protecting French Protestants from persecution. Religious wars continued to flare up across Europe, but more and more enlightened rulers stopped listening to church leaders. During the seventeenth and eighteenth centuries, philosophers speaking of universal rights of mankind, and political theorists designing international peace based on treaties instead of fleets, dared to envision a new world order. By the eighteenth century, it became practically impossible for the churches to conduct inquisitions with unlimited force and order executions for heresies against their own creeds. A new idea, a philosophical and political idea, had become the new European standard of civilized peoples: the freedom to practice one's own religion, as a citizen under a government protecting the same for rights for everyone.

Over 350 years after the Edict of Nantes, the Roman Catholic Church at last renounced what had long been its God-given right to oppress, torture, and murder heretics. The Declaration on Religious Freedom, Pope Paul VI's *Dignitatis Humanae* (December 7, 1965), admitted what the rest of the civilized world had understood for a long time. Its opening paragraph amounts to a confession that the Church has learned from the modernizing world something it never knew about the human spirit.

> A sense of the dignity of the human person has been impressing itself more and more deeply on the consciousness of contemporary man, and the demand is increasingly made that men should act on their own judgment, enjoying and making use of a responsible freedom, not driven by coercion but motivated by a sense of duty. The demand is likewise made that constitutional limits should be set to the powers of government, in order that there may be no encroachment on the rightful freedom of the person and of associations. This demand for freedom in human society chiefly regards the quest for the values

proper to the human spirit. It regards, in the first place, the free exercise of religion in society. This Vatican Council takes careful note of these desires in the minds of men. It proposes to declare them to be greatly in accord with truth and justice.[5]

Today, there are fortunately just a small number of countries where the dominant religion has the kind of immense power over society and government enjoyed by the medieval Church. In every one of those countries, the dominant religion continues to be oppressive, malevolent, terrorizing, and genocidal. Many of those countries happen to be controlled by Islam. This is a regrettable but changeable situation. Islam, like Christianity (and Hinduism, Buddhism, etc.) can take more benign forms where a strong government counterbalances religious faith. It is false to claim that any religion must always be oppressive.

Evidently, religion is quite capable of domestication into peaceful forms where a strong nonreligious government is responsible for law. But the history of the world also demonstrates that a religious government invariably tends to abandon good will towards humanity and instead discovers that its God wants submission at all costs.

Religions often throw this challenge at the nonreligious: "Show us your perfect ethical system to replace ours, if you judge religion immoral." This challenge is just a bluff and a distraction. No "perfect" ethical system is needed to notice when a powerful religion abandons basic moral virtues and descends into depravity. Common sense should be enough for boldly declaring that all people have the right to freedoms of conscience, belief, and association. Revolting against rule by priests and kings "appointed" by God, and founding a government elected by citizens, only requires confidence in oneself and faith in one's neighbors. No religion can take credit for inventing the ideas of human rights or mass democracy.

If some religions today now claim that God approves of human rights and popular democracy, we can applaud their acknowledgement of enlightened reason, and hope that the rest of the world's religions awaken as well. However the future unfolds, we can never forget that only secular government ensures that religions remain domesticated.

Notes

1. Singer, Isidore, and Cyrus Adler. *The Jewish Encyclopedia: A Descriptive Record of the History*, vol. 12. New York and London: Funk and Wagnalls, 1906.

2. Hesiod, *Works and Days*, in *The Homeric Hymns and Homerica*, trans. Hugh G. Evelyn-White (Cambridge, Mass.: Harvard University Press; London, William Heinemann, 1914), lines 276–284.

3. Herodotus, *The Histories*, trans. A. D. Godley (Cambridge, Mass.: Harvard University Press. 1920), book 3, chap. 38.

4. W. R. Inge, *Christian Ethics and Modern Problems* (London: Hodder & Stoughton, 1930), pp. 182–183.

5. http://www.vatican.va/archive/hist_councils/ii_vatican_council/documents/vat-ii_decl_19651207_dignitatis-humanae_en.html

Further Reading

Assmann, Jan. *The Price of Monotheism*. Stanford, Cal.: Stanford University Press, 2010.

Baker-Brian, Nicholas. *Manichaeism: An Ancient Faith Rediscovered.* London and New York: T. & T. Clark, 2011.

Bloom, Paul. *Just Babies: The Origins of Good and Evil*. New York: Random House, 2013.

Broom, Donald. *The Evolution of Morality and Religion*. Cambridge, UK: Cambridge University Press, 2003.

Cole, Phillip. *The Myth of Evil: Demonizing the Enemy*. Westport, Conn.: Greenwood, 2006.

Frances, Bryan. *Gratuitous Suffering and the Problem of Evil*. London and New York: Routledge, 2013.

Green, Toby. *Inquisition: The Reign of Fear*. New York: Macmillan, 2009.

Harris, Sam. *The Moral Landscape: How Science Can Determine Human Values*. New York: Free Press, 2010.

Hunt, Robert P., and Kenneth L. Grasso. *Catholicism and Religious Freedom: Contemporary Reflections on Vatican II's Declaration on Religious Liberty*. Lanham, Md.: Rowman & Littlefield, 2006.

Janney, Rebecca Price. *Who Goes There? A Cultural History of Heaven and Hell*. Chicago: Moody Publishers, 2009.

Miller, Sukie. *After Death: How People around the World Map the Journey after Life*. New York: Simon and Schuster, 1998.

Pagels, Elaine. *The Origin of Satan: How Christians Demonized Jews, Pagans, and Heretics*. New York: Random House, 2011.

Rose, Carol. *Giants, Monsters, and Dragons: An Encyclopedia of Folklore, Legend, and Myth*. New York: W. W. Norton, 2000.

Shermer, Michael. *The Science of Good and Evil: Why People Cheat, Gossip, Care, Share, and Follow the Golden Rule*. New York: Macmillan, 2005.

Smith, Mark S. *The Origins of Biblical Monotheism: Israel's Polytheistic Background and the Ugaritic Texts*. Oxford: Oxford University Press, 2001.

Stoyanov, Yuri. *The Other God: Dualist Religions from Antiquity to the Cathar Heresy*. New Haven, Conn.: Yale University Press, 2000.

Turner, Alice K. *The History of Hell*. Boston: Houghton Mifflin Harcourt, 1993.

7

LIFE AND DEATH

From birth to death, mythologies have placed vast religious significance upon the primary transitions of human life: how a person enters the world, and how a person leaves it. Religious perspectives on these transitions take quite different forms. From embryology, birth, and the status of infants to the meaning of illness, dying, and death, the world's religions have taken all sorts of stances on these vital matters in fascinatingly divergent ways. Issues such as pregnancy and birth, abortion, health and sickness, old age, dying, and the definition of death have also engaged science's attention. The scientific worldview provides understandings of all these human transitions which often disagree with religion's presumptions and preferences.

Pregnancy and Birth

Religions around the world have long regarded pregnancy and birth of a baby as infused with spiritual importance and religious significance for the entire social group. The world's religions have views about the moral value of the fetus and baby, which sometimes influence the laws of nations.

Hinduism has a long tradition of moral objection to abortion and infanticide, going back to several of its earliest scriptures. Passages in the Rig-Veda, along with the Laws of Manu, urge protection of the fetus and condemn killing fetuses and babies. Other Hindu sources of ethical guidance provide additional details about the rare permissibility of abortion, such as situations where the health of the mother is a concern, the pregnancy is proceeding abnormally, or the fetus is not growing properly. The long history of medical practice in India shows that physicians were able to induce abortion on occasions that called for that procedure, but it was not a morally approved practice. Modern India has gradually accepted widespread access to birth control and abortion, but the Hindu antipathy towards abortion means that it is a private matter not to be discussed outside the family. Abortion is frequently used in India today to have more boy babies than girl babies, but this is not new in India; historical texts reveal that abortion has been used to prevent girl births for many centuries.

Buddhism, like Hinduism, has regarded abortion as morally questionable at best, and something to be avoided during later months of pregnancy. Buddhism has never had a uniform doctrine regarding the moral value of the fetus, or any clear consensus about the timing of the stage when a fetus becomes fully human. The ancient Buddhist texts from the earliest centuries of Buddhism make no specific declarations about these matters, and they do not express any prohibition against abortion. However, Buddhism has traditionally disfavored abortion, because of its prohibitions against killing living beings, and its acceptance of the idea of the transmigration of souls. While Buddhist theology specifically denies the Hindu conception of a substantial personal self that survives death and reimplants in a new human life, Buddhism does agree that everyone who dies without attaining the final oneness with Nirvana must undergo spiritual rebirth in a new body. As Buddhism spread from India, it carried along a preference for the protection of all human life, including the fetus. Most Buddhist countries today display a moral aversion against abortion, but they have few laws restricting access to abortion. Some branches of Mahayana Buddhism that developed in southeast Asia did later arrive at the belief that a reincarnated soul joins with its new embryonic body soon after the moment of conception, but this doctrine isn't accepted beyond those branches of Buddhism. The most widely shared conviction about abortion across Buddhism is that any consideration of abortion must be undertaken with the duty of compassion firmly in mind. Compassion for the pregnant mother as well as compassion for the fetus must be taken into account, and abortions done for the sake of the mother's situation are still regrettable killings.

China had no laws regulating abortion until the late Qing dynasty in the nineteenth century, under the influences of Christianity and Western law. By the 1960s, legal prohibitions against abortion were relaxed, and China's one child policy brought a period of relatively easy and legal access to abortion. The current legality of abortion in China cannot be taken to imply that the Chinese culture has no moral problems with abortion. Quite the opposite is the case. However, traditional Chinese views about pregnancy and the fetus did not rule out abortion, either. The oldest religious traditions in China, Daoism and Confucianism, did not specifically supply opinions about the moral status of fetuses. Pregnancy, like all other bodily functions, was regarded as a natural process that should be permitted to come to its natural conclusion. Whether a pregnancy turns out well or badly depends almost entirely on nature's own ways, or on "the will of Heaven." When pregnancies go badly and the mother's health is suffering, ancient Chinese doctors did have a variety of crude techniques and drugs to cause a miscarriage, but whatever the outcome, responsibility was mostly placed on fate.

Buddhism's arrival from the West into China influenced medicine. Many medical texts dating back to the sixth century discourage healers from participating in unnecessary abortions or giving poisons to induce abortion during the later months of pregnancies. The great physician Sun Szu-miao (c.581–682) wrote the most influential treatise on medical practice in China, titled *Qianjin Yaofang*. The first chapter of this great work concerns the moral principles and sound wisdom of practicing the healing arts. There is no mention of forbidding participation in abortion. However, the influence of this treatise was generally conservative, discouraging medical healers from involvement with unnecessary abortions. Chinese Buddhism did not contradict the older traditional understanding of the fetus's development as having a spiritual aspect as well as a physiological aspect, and these two features grow together during pregnancy. Over the centuries, medicine in China was able to study fetal development. None of the three major Chinese religions thought that the soul or spirit is present during the earliest months of fetal life. They also agreed that by the stage of late development, the fetus was a human being with its own spiritual nature. Buddhism had the least direct impact in Japan, where traditional Shinto belief has held that the fetus did not acquire a spirit until the event of birth.

In the West, the Greeks commonly practiced abortion and infanticide. The Hippocratic Oath, which includes a vow to abstain from giving drugs to cause an abortion, was not an ethical creed common to medical practice in Greece. Only those healers belonging to that Hippocratic school of medicine were urged to conform to that oath, and this particular school was never very large, nor was it representative of most Greek physicians. The Hippocratic school was unusual in several ways, keeping it out of the mainstream Greek opinion. The only other school of thought that similarly prohibited the killing of any human being was the older Pythagorean tradition, which similarly regarded the fetus as fully and spiritually human from conception and announced its aversion to another common Greek practice, euthanasia for those hopelessly suffering in old age. The Hippocratic Oath would have remained a curiosity of ancient Greek medicine except for the rise of Christian ethics centuries later. Early Christian thinkers who appreciated the value of Greek philosophy endorsed Hippocratic medicine, ensuring that Hippocratic texts survived in libraries, but nowhere does an early Christian theologian endorse the Hippocratic Oath. The high value of the body of preserved texts about health and medicine from this Hippocratic school meant that they went on to survive the Dark Ages and were extensively copied and recopied in Medieval Europe, supplying much of the foundation for academic medicine by the twelfth century.

Only by that time did the Hippocratic Oath receive much attention from Christian theologians, and they had to modify it to eliminate references to Greek gods, cultish pagan ideas, and vows of loyalty to one's teachers. The Christian version of the Oath also strengthened the prohibition of abortion, forbidding abortion after "quickening," the time when the fetus could be felt moving, around the fourth month of pregnancy. Greek philosophers such as Aristotle had already regarded quickening as the time when the fetus developed its own mind. Church theologians such as St. Augustine and St. Anselm agreed, and this view was affirmed by the 1312 Catholic Council of Vienna. Islam was also familiar with Aristotle and the body of Hippocratic texts on medicine, and Islamic versions of the Oath were sporadically used in the Middles Ages. Islam continued to regard abortion as permissible and legal before the fourth month of pregnancy, down to the present day.

By the time of the Renaissance, the Hippocratic Oath and its prohibition against abortion had almost no detectible role in academic medicine or theology, and the Oath was not a feature of academic ceremonies. However, the occasional mention of the Oath in European writings during the fifteenth, sixteenth, and seventeenth centuries showed that its Christian form was acknowledged as a valid ethical code for physicians, although no one was publicly affirming it. Theological ethics was so far advanced by that era that a crude oath of simplistic maxims couldn't be taken seriously as the last word on the duties of a physician. And neither euthanasia nor abortion were significant ethical issues for either the Catholic or Protestant churches, until the mid-nineteenth century. European countries did not have legal codes forbidding or punishing abortion until the nineteenth century. In America, the law had continued to follow the British Common Law rule that abortion should only be illegal after "quickening," the time when the fetus could be felt moving, around the fourth month of pregnancy.

The Hippocratic Oath was revived in America in the late nineteenth century in a further modified form, coinciding with a growing political trend towards making abortion illegal in both Europe and America. Before World War II, only a small percentage of medical colleges, mostly associated with religious denominations (especially Catholicism), were requiring an oath that closely resembled the original Hippocratic Oath. After World War II, most medical schools required a thin version of the Hippocratic Oath, but most versions (like the one endorsed by the American Medical Association) omit mention of abortion and euthanasia.

The nineteenth century was a time of dramatic change in religious attitudes towards abortion. Although a few prominent theologians of the early and medieval Church thought that the fetus has a soul from the beginning of pregnancy, most continued to agree that the soul joins the body at quickening. The Roman Catholic Church did not explicitly condemn all abortion and promulgate the doctrine that conception is the moment when the soul joins the body until 1869 with Pope Pius IX—the same Pope who declared himself infallible on doctrinal matters—and his declaration "Apostolicae Sedis Moderationi." Unable to appeal to any clear statements in the Bible about fetuses or the immorality of abortion, Pius IX relied mostly on speculative theology and metaphysics of his own design to invent this "infallible" conclusion.

As for the Protestant Churches, they also have an uncertain and shifting record on the fetus and abortion. Both John Calvin and Martin Luther, the primary founders of sixteenth-century Protestantism, thought that human life began at conception, but the major Protestant denominations did not impose this view strongly on their members or fight to enshrine these dogmas in legal restrictions. Most of Europe continued to regard abortion as permissible before the fourth month, and when the health of the mother demanded medical action. In America, Protestant denominations morally condemned most abortions during the late nineteenth and twentieth centuries, yet most of them supported a woman's right to obtain an abortion by the early 1970s, and announced their endorsement of the *Roe v. Wade* Supreme Court decision in 1973. Only later, by the 1980s, did an alliance emerge between the Catholic Church and evangelical churches, especially the Southern Baptists. This politically powerful alliance has prioritized the moral condemnation of abortion, from the start of pregnancy, without exceptions such as a case of rape or to save the life of the mother.

Longevity and Immortality

Recommendations for living well in old age is a frequent theme in religious literature, since the earliest recorded mythologies. Enjoying health despite great age is taken as a sign of living life according to the highest virtues, and probably according to divine expectations, as well. The godliest are the healthiest.

According to Jewish myth, the first man, Adam, lived to be 930 years old. But that's not the Jewish record. His descendant Methuselah reached the greatest age listed in the Bible, 969 years of age. And Methusalah's grandson Noah shared in this amazing longevity. Noah had his three sons after he reached the age of 500 years, and survived the flood to live to the great age of 950. But the Bible goes on to record briefer ages for Noah's descendants, as successive generations down to Moses lived shorter lives. Moses himself received only 120 years.

If the Jewish priests were trying to keep up with neighboring religious myths, they did a modest job. According to the Sumerian myths dating from two millennia before Judaism, the first rulers during a golden age of humanity lived for at least 18,000 years, and some reached over 30,000 years of age. That truly is longevity. Alas, great longevity was only a temporary thing. The Sumerians were the first to tell a legend about a great flood destroying most of humanity, and they similarly blamed the flood for reducing human longevity afterwards. The great kings of the next age of the world only lived around a thousand years. From the Mediterranean to Japan, the mythologies of the early civilizations all agreed that their ancestors, among the first to walk the earth, were somehow closer to the divine creative energies that made the world, and closer to the gods themselves.

In the beginning, these ancient myths agree, there was a "golden age" in which "the wise rulers" were practically demigods capable of prodigious accomplishments, profound wisdom, righteous deeds, and, not surprisingly, living extremely long lives. Yet that golden age could not last. The ancient Hindu scriptures describe cyclical ages, starting with a Golden Age of highest virtue and greatest health, but that age was long ago, and today's world is a dramatically poorer age. Mythic legends about an early period of vast prosperity and abundance were proliferating across the ancient world. These legends reached the ancient Greeks, and the poet Hesiod organized his own classification for the age of the world in his *Works and Days*: the present age is an "Iron Age" that must be far worse than the preceding "Heroic Age," but best of all was an initial "Golden Age" when humanity lived like gods.

Many indigenous religions similarly claim that their ancestors had lived to extreme old ages lasting many centuries, but they admit that people can't reach such extreme old ages anymore. Still, religions remain capable of promoting the idea that religious purity can produce impressive longevity. The oldest tribal chief, village elder, or expert healer are still assigned impressive ages; it isn't unusual to hear about a well-respected elder or religious hermit in Central Africa, or Pakistan, or China who is locally believed to be 120, 150, or 200 years old. In rural China, Daoist legends attribute extreme old age to expert practitioners of the ancient ways of diet and exercise. In Chinese traditions, aging is entirely natural, and hence there must be natural remedies to forestall aging. People in rural India also routinely accept claims that Hindu yogis and mystics have attained ages of 200 years or more. Fascinated with any opportunity to enjoy health in old age, and live out many years far beyond the normal life span, many religious people are consistently eager to believe in, and practice themselves, a strenuous or purified way of life. The more ancient the wisdom, the popular thinking goes, the most trustworthy and practical, since methods of reaching human excellence and longevity, and perhaps even immortality, were understood by the worthy ancestors. Human perfection lies in the distant past—only by closely imitating ways of life from that earliest age could anyone achieve similar healthy results.

We should invent iron. If we had iron,
civilization wouldn't be so prone to collapse

Not without a degree of irony do historians continue to speak of the "Iron Age." The actual Iron Age is simply a convenient label for the rise of cultures capable of making and using iron for tools and weapons, which began around 1200 BCE. The rise of iron use was fairly gradual, but that 1200 BCE is convenient for demarcating the abrupt end of the Bronze Age, a time period when almost every advanced culture from Greece and Egypt all the way east to Persia and India suffered from extreme turmoil, terrible war, or complete political collapse. Even China suffered along with the rest of civilization, as the Shang dynasty collapsed in 1046 BCE and a long period of Chinese disunification followed.

This Bronze Age "collapse," acknowledged by academic historians as the greatest widespread retreat of human civilization ever recorded, was followed by Iron Age lamentations over the terrible losses. It cannot be a coincidence that a new kind of religious literature arose after 1000 BCE, the mythologies that describe a "Golden Age" of human excellence and prosperity enjoyed by distant ancestors but gone forever. In India and China, religions taught that the only kind of longevity possible anymore consist of some sort of reincarnation into new bodies. Perhaps your indestructible life energies will be reused by more earthly life after your death (China), or your soul will leave your body and enter a newborn life here on earth (India). In the Middle East, new religions taught that each person's soul will have second spiritual "afterlife." This Iron Age innovation, the radically novel idea that each and every person is guaranteed another life, no matter who they are, enjoyed vast appeal among peoples who knew that the natural prosperity of their ancestors could never be theirs.

Some sort of destined immortality, available to each individual, played a huge role in Iron Age civilizations to an extent not observed in any Bronze Age cultures. The idea that immortality is available to some special individuals is hardly an Iron Age idea—from the beginning of recorded scriptures from the earliest civilizations, special demigods, rulers, heroes, and spiritual figures get a chance at immortality. The early Pharaohs and priests of Egypt were obsessed with ensuring their immortality through mummifications and pyramid schemes. The Sumerians had Gilgamesh, the fifth king of Uruk in Mesopotamia (probably an actual king, dating around 2500 BCE) who, according to legend, achieved demigod status and played the starring role in the Sumerian *Epic of Gilgamesh*. Although Gilgamesh fails in his quest to find a way for humans to achieve immortality (the needed miraculous plant is taken by a serpent instead), the core religious notion remains: special heroic figures can potentially reach immortality.

By the time Christianity arose, a very different understanding of immortality pervaded the Mediterranean, the Middle East, and India. Neither longevity nor earthly immortality is the religious goal anymore. A new notion of immortality was now in control: each person is guaranteed a long afterlife (hopefully a pleasant afterlife) to come, after a regrettably short and turbulent mortal life. There is little point to attempting extraordinary natural vitality in order to prolong this earthly life as much as possible. The sages and hermits who seek natural longevity are truly special people, each religion from Christianity and Islam to Hinduism and Buddhism all still agree, but their lonely path is not for the common people. Instead, only virtue and righteousness despite life's disappointments and troubles could supply the path towards the only kind of longevity still available: one's spiritual afterlife. Only one's harmony with divine powers, and not with nature's ways, could relieve people from their mortal sufferings.

Disease and Dying

Lacking any knowledge about the genuine physical causes responsible for most diseases, nonscientific peoples have relied on religious ideas to explain and deal with illness, infirmity, and dying. An exhaustive study of the world's religions, past and present, would be necessary to discover the very small number of religions that don't rely on malevolent spirits to help explain human misfortune and suffering.

The general correlation is not hard to see: the older a religion, the greater reliance on threatening spirits responsible for illness. The shaman or "medicine man" of an indigenous culture possessed a modest body of knowledge about dietary or herbal remedies for ordinary ailments, along with a vast body of lore and legend about rituals necessary for avoiding or alleviating sickness and death. The reason why indigenous lore about health and sickness relies so heavily on ideas about sustaining good social relationships, especially relationships with potentially dangerous spirits, is simply because no indigenous healer could know much about the physiological causes of disease. A well-intentioned healer ultimately only had people to work with, and could only arouse whatever healing energies could be generated in interpersonal relationships.

The indigenous healer played the role of counselor and mediator, working at a group psychology level, not a physiological level. As a mediator among the people, and a mediator between the people and the spirit world, the healer was working with the only materials available to make a difference. The ultimate source of the problem responsible for illness and death really wasn't within the sick person's bodily organs. The indigenous healer knew almost nothing about disease and physiology. The ultimate problem therefore had to do with interactions between living beings: people with each other, people with nature, or people with an evil spirit.

The spirit world says take two leeches
and call them in the morning

When advocates for traditional healing methods justify prehistoric rituals, they typically resort to praising the intensely social and psychological powers of group engagement and mutual concern. This manner of defending traditional ways is quite understandable. What else could be said in favor of indigenous practices? Lori Arviso Alvord, who describes herself as both a Navajo Indian and the first member of her tribe to become a surgeon, declares herself unable to connect hardly anything from her tribe's heritage with modern medical practice. She writes, "During my training as a surgeon, I was unable to harmonize my background as a Navajo with my medical world."[1] This could not really be surprising. And Alvord easily explains how traditional healing requires a healing social environment (rather than offering an individual physiological remedy).

> Ceremonies reinforce the belief that we live in harmony with the animal world and the natural world. Humans value many things, but they often assign greatest value to family, or that which they consider sacred. Many Native American tribes have assigned both a spiritual and a familial value to the animal world and the environment. The earth is "mother," the sky is "father." The eagle and bear are "brother." Mother Earth is sacred in her mountains and valleys. The relationship of humans and their environment is one of deep respect, a desire to protect and defend the animal world and the environment. The protective element provided by spirituality has direct healing effects on human beings. By keeping the environment protected, we have clean air to breathe, clean water to drink, clean earth in which to grow plants. We are shielded from the illness that results when our world becomes toxified.[2]

Alvord goes on to point out how community mindedness may be responsible not only for preventative methods, but also for genuinely healing powers as well. Positive social relationships and communal ritual can arouse helpful physiological responses from many bodily systems.

> Ceremonies are often performed for the purposes of healing. Many of the forces of healing used in ceremonies have already been described.

The effects of stress and depression on the immune system are better understood, and the effects of ceremonies are easily understood in this context. These principles are now beginning to be used by other healing systems as well. Western medicine is waking up. It has started to realize the power of healing that exists beyond the realms of procedures and medications. Studies have started to prove the healing power of such realms as support group therapy, music therapy, healing and the arts, spirituality and medicine, pet therapy, massage therapy, and so on. We are learning that healing can be influenced by multiple forces within our lives, that we are deeply interconnected to all aspects of our lives, and that we can immerse ourselves in many areas to achieve healing.[3]

Defenders of traditional healing practices, from the Americas to Africa and across the breadth of Asia, can point to helpful aspects of communal thinking. If strengthened communality among people were all that is involved, the effects of indigenous practices could shine in a fairly positive light. However, indigenous religion always adds unnatural spiritual beings to the equation. Where indigenous religion has large control over a society, negotiating with those spiritual beings can take priority over utilizing all available practices, traditional as well as medical.

Cataloguing all the ways to negotiate with the spirit world, undertaken by religions on every continent, requires vast ethnographic research. Religions are able to convince people to transfer property and power to special human practitioners skilled at obtaining those spiritual favors on behalf of their clients. As a way to establish a permanent kind of profession, a profession resting on nothing more than compelling stories, nothing rivals religion. Most authors of literature find it difficult to sustain a career on writing alone, no matter their genius; a charismatic and convincing storyteller about the spirit world never goes hungry.

There are endlessly complicated ways to persuade people that they urgently need information, protections, or interventions from one spirit or another. When the possibility that other people themselves may possess spiritual powers capable of affecting others is included, the permutations and combinations of religious activities inflate to vast proportions. Without losing the ability to recognize such complexities, scholars lump together sets of religious practices under convenient labels, even if they sound outdated to modern ears. For example, determining potent spirit forces affecting one's life amounts to "divination," and malevolent individuals wielding harmful spiritual powers might practice "witchcraft," as an article in the *Wiley-Blackwell Companion to African Religions* recounts:

> Witchcraft is deeply embedded in African traditional systems of belief and practice and remains formidably ensconced in daily consciousness even in the contemporary urban situation. It is conceived to be a very real, prevalent, and menacing force. The perceived need for protection from witchcraft is a decidedly compelling factor in seeking refuge in traditional rituals such as divination and its prescriptions for purification, sacrifice, protective amulets, and herbal remedies.[4]

If spiritualist healers were only supplementing genuine medical remedies, the dangers of indigenous religion could be minimal. However, where indigenous religion maintains control over social convictions, so-called healers can instead operate their well-paying practices in competition with scientific medicine.

There are far too many tragic examples of local religions obstructing public health and raising death counts to keep track. Africa again supplies a present-day tragedy: reducing the spread of HIV/AIDS in many African countries is obstructed by locally trusted spiritualists who divert people's attention away from preventative measures and efficacious medicines and towards their personal advice and home-made remedies, sold at a nice profit. Preserving traditional cultures and indigenous healing knowledge in the face of pervasive colonialism can be noble work, unless millions are dying in the process. Rightly faulting centuries of European conquest and control for the breakdown of traditional culture doesn't justify excusing the remnants of indigenous spiritualism from their involvement in perpetuating harmful superstitious ignorance. Many countries around the world, including modernizing countries such as India and China, confront traditional practices that claim to have the real power over health, life, and death. Religions taking advantage of ignorance by sustaining fascination with spiritual forces that supposedly control life and death are a feature of every historical epoch and geographical region.

Besides helpful spirits (like angels) or dangerous spirits (like demons), a widely feared kind of spirit is the ghost, the sort of unnatural thing that blurs the otherwise sharp line between life and death. By whatever name a culture calls it, a lingering spirit, once enlivening a human body, is a fearful thing to have still roaming this earthly world. Ghosts can do far more than frighten people, according to many religions. They can cause trouble for the family, bring "accidents" or illnesses to those they haunt, and generally arouse deep anxieties in those who must deal with a ghost. Belief in things akin to ghosts is extremely common across societies having an indigenous religion, along with rural areas where Christianity, Hinduism, or Chinese religion prevails. About a third of Americans today believe in ghosts.

Belief in ghosts is connected with religious beliefs about the appropriate destiny of the soul after death, and the right ways that one's soul should be freed from the body during the dying process. For example, it is a common notion that ghosts are the troubled spirits of people who died in tragic or disturbing ways, or that they had themselves committed horrible deeds. All religions agree that ghosts should not be still here among us, but something or someone is preventing their release to leave this world and go to the correct place for souls. Religions usually recommend "best" ways of dying, if the timing and manner of death can be at all controlled. The process of dying and the moment of death is therefore charged with great significance—the fate of the soul after death could be at stake. Dying in a peaceful and natural way would be ideal, so the soul can be focused on the right sort of transition to the next world. That's the obvious answer. When it comes to "unnatural deaths," being killed isn't preferred, and most religions past and present regard unexpected death at the hand of another person as an untimely misfortune.

Religions around the world are now confronted with the issue of euthanasia in more pressing ways, due to scientific medicine. Poorly understood poisons, such as the hemlock that Socrates drank, have been used to provide deadly relief towards the end of life long before anything like scientific medicine was established. It's not scientific medicine that first raised this moral issue, but the new technologies capable of keeping the body alive longer and longer are now intertwined with ancient concerns. These moral issues make it necessary to ponder what death really is, and how it really happens. Is euthanasia an unnatural death to be avoided, or would euthanasia be a natural way to recognize the limitations of life?

Where a kind of euthanasia involves killing, such as the deliberate killing of a patient in a medical setting, most religions raise staunch objections. If a way of helping a person to die needn't be understood as a killing, such as the surrender of a patient's life to natural causes of death, then euthanasia can be regarded as much more acceptable. For example, Christianity widely supposes that there is no duty to continue pointless medical attention when death is inevitable or the patient cannot possibly benefit from further treatment. Even Catholic theology permits ending medical treatments, allowing a patient to finally die, if no benefit is gained from merely sustaining bodily life. Between firm moral prohibitions against killing, and dogmatic ideas about letting God's final plans be fulfilled for us all, religions tend to take strong stances against most forms of euthanasia. Islam, Hinduism, and Buddhism share Christianity's stern distrust of euthanasia. In China, there is relatively less traditional resistance to euthanasia, although it is officially discouraged and generally illegal. Chinese religions, including Daoism and Confucianism, do not promote a theological position sharply distinguishing the body from the spirit, and they do not worship a theistic god who commands obedience or controls fate.

The Soul and Death

Taking a nonreligious position on death and dying, medical science views life as a process, so it views dying and death as processes as well. The notion of "the moment of death" has intuitive appeal to people ignorant of biology. Popular entertainment reinforces this notion, to create a climactic scene about a character who must die, but not before uttering their final words for the camera. This is highly effective drama. Our emotions, and not our smarts, is placed in control over our moral intuitions.

Religious-minded scientists haven't been immune to this popular notion. The famous experiment from 1907 by Massachusetts doctor Duncan MacDougall, who tried to measure the weight of the soul, illustrates how religion and science have been blended together here. If there is a moment of death, MacDougall reasoned, and if the soul has physical qualities as well as spiritual qualities, then the most fundamental physical quality, mass, should be involved. When a person dies, MacDougall concluded, the body should be measurably lighter—and his terminally ill patients (six in all) consented to his experiment to test this hypothesis. They were hoisted onto an industrial capacity scale,

21 Grams?!? I've lost weight!

and MacDougall was able to watch their dying breath and then turn to watch the scale's pointer. MacDougall claimed that some of the deaths were accompanied by lighter weights according to the scales, and an average of 21 grams was published.

The facts that an industrial scale designed to weigh masses up to hundreds of pounds cannot reliably make a measurement in grams, and that any small change in perspective looking at a large scale pointer can make it look like the pointer had truly moved, means that MacDougall's experiment is scientifically meaningless, despite the way that his "evidence" for the soul has been widely repeated ever since. It is also intriguing that MacDougall found no "soul weight" when dogs died, and his dying sheep actually appeared to gain weight for a time.

No scientific experiment has ever revealed the slightest amount of serious evidence for anything spiritual connecting to a physical body. This lack of evidence is not because science is necessarily close-minded and hostile towards the notion of the unnatural. Science discovers "unnatural" things all the time, in the sense that novel phenomena are discovered in fresh experiments that don't fit into the established scientific worldview of what counts as natural. That's when creative scientific theories try to figure out how these unexpected phenomena can fit with what is known about the natural world. The quantum behavior of subatomic events couldn't have been more unnatural, according to what counted as "natural" at the end of the nineteenth century—for example, things behaving like particles and then like waves, and particles having instantaneous connections despite a vast distance between them. All the same, physicists have been able to show how laws of nature continue to prevail over all these phenomena, and how the laws of the quantum world work with the laws of the ordinary world. Nature has gotten more complicated in the process, but science isn't discovering the "spiritual" along the way.

Despite these impressive scientific achievements, plenty of religious enthusiasts ignore the real science and instead propose that the spiritual world has been revealed in the laboratory. Is science really helping to understand the spiritual life? Instead of reducing mystery, asking such questions only manages to combine and multiply mysteries. The longstanding mystery of a nonphysical soul having any connection or interaction with a physical body will remain a stubborn mystery, simply because neither reason nor common sense could ever find a way for two things to relate when they share nothing in common at all. Undeterred by mystery, anything that seems mysterious strikes the religious mind as the very key to understanding. Better yet, two mysterious things might be actually related to each other. Fascination with anything involving the quantum world is sweeping through the audience of spiritual-minded people nowadays—for example, what is really behind anyone imagining that a person's "spiritual" consciousness comes from the quantum realm? Religions have long been in the business of attributing religious significance to surprising new science, ever since ancient Babylonian times when the regular motions of the stars looked to religious thinkers like heavenly tracks guided by a divine mind. As astronomy found the cause for stellar and planetary motion in mindless forces like gravity, divine mind was eliminated along with mystery.

If religion relies on medical science to help determine how the spirit leaves the body, it will be disappointed yet again. Modern science can't agree with any religious ideas about death, because contemporary physiology knows so much more about life.

Let's start with breath. Breathing is crucial to life, no doubt. The English word "spirit" comes from the Latin language, where "spiritus" originally meant "breath" before it was borrowed for expressing a religious idea. The Latin word for soul is "anima," which also supplies the root word for "animated" and "animal," and it also originally meant "breath." The Greeks had their own word for spirit—"pneuma"—which also refers to breathing, as well as "air," and the Greeks could also talk about the mind using their word "psyche" (yielding the label of "psychology" for the science of the mind), but "psyche" also traces back to an early Indo-European word for breath. Other languages from those Indo-European family roots, such as the Scandinavian languages and the Baltic and Slavic languages, use a word meaning "breath" for "spirit" or "soul." "Breath" is also the root meaning behind the concept of spirit or soul in Semitic languages such as Hebrew and Arabic, the Sanskrit words for "life" (prana) and "self" (atman), and even the Chinese word for spirit, "qi," means "breath." And many more languages around the world connect the concept of spirit or soul with breath, as well as air and wind. There is little need for speculation about such a massive coincidence; it's no coincidence, since anyone can observe how death involves the cessation of breathing the air. So long as the body is breathing, death has not yet come. And when breath has left the body, the spirit has left. Just a little imagination was needed to suppose that the spirit of life is made of air, or something very much like air—having no weight or specific size, able to pass in and out of more substantial materials, invisible to the eye and ungraspable by the hand.

Given this closest of connections between spirited life and breathing, myths about gods and humans frequently credit a god for bringing people to life by infusing them with breath. Just a sampling from around the world is possible here. The Malagasy tribe of Madagascar relate a myth in which little clay dolls come to life to become the first people after the supreme creator god breathes life into them. The Aztec god Quetzalcoatl creates wind and the breath of life. The same role is played by the American Indian god of the Creeks, Esaugetuh Emissee, a name meaning Master of Breath, who formed the first humans from clay and breathed life into them. The ancient Egyptians, thousands of years before Christ, regarded their supreme gods Amun and Re as the gods responsible for supplying the "breath of life" to living creatures made of dirt. Other Egyptian gods could supply breath too—at one climactic point in the myth of the gods Isis and Osiris, Isis is able to bring Osiris back to life by forcing the breath of life into his corpse. The Sumerian god Enki successfully brought clay forms to life by breathing life into them, creating the first humans. Surrounded on all sides by older civilizations and their myths about gods breathing life into the first humans, the Hebrews included this myth in Genesis 2:7, "And the Lord God formed man of the dust of the ground, and breathed into his nostrils the breath of life; and man became a living soul."

When the body breathes no more, death has come. But what causes breathing to stop? Ordinary observation tells anyone that the body requires other energies besides airy breath: the flow of blood is necessary, and so is food. People can live without food for quite a while, but the way that serious loss of blood, and stoppage of the heartbeat, will soon cause breath to depart the body, was something that could not go unnoticed. The close relationship between breathing and blood flow was suspected by Greek thinkers, and Galen's treatises (composed around 180–200 CE) agreed that the lungs help convey the life energy in breath into the blood. Not until the late Renaissance could experimenting doctors finally track how the heart and lungs work together to transfer air's energy through the blood in arteries to the rest of the body. And not until the 1800s could biologists figure out that oxygen was the crucial element in the air, which combined with digested food to extract energy for life.

Religions were not wrong to associate life with breath, since that link was common knowledge for anyone long before mythic storytellers got involved. But religions have been entirely wrong to think that anything spiritual, airlike or not, is really involved with life. The real story is more fascinating than anything imagined by a religious mind. Let's start with breathing. First of all, there's no energy in the air to be extracted for life. Air is mostly composed of nitrogen, which is a vital element for plants (it is a typical component of plant fertilizer), but animals get nitrogen by eating food, not breathing air. What animals need is the oxygen in the air, but the lungs are merely a transfer point, where red blood cells absorb oxygen and carry it to the body's cells. Circulating blood also carries the other digested molecules needed for life—mostly chemical enzymes, proteins, and sugars. That's where all that oxygen does its work. The real breathing of metabolic respiration happens inside all the cells of your body, not inside your lungs.

Many kinds of nutrients supply cellular "building blocks" for regenerating tissue structures, but the energy that cells need to get work done is locked inside those sugars, mostly in the simple form of glucose, $C_6H_{12}O_6$. The energy cells need from that molecule is locked in the chemical bonds between the atoms of those three elements. How did that energy get trapped there? That glucose molecule was forged in the cell of a plant exposed to light. A photon from the sun got caught by an electrochemical "machine" in the chlorophyll of a plant cell, where its solar energy got converted into an electron's energy and then into a chemical bond involving that electron to permit a sugar molecule to be assembled from carbon dioxide and water in the cell. The energy inside glucose is the energy of starlight, and an animal cell unlocks that energy by adding more oxygen. To think that your tiny cells are operating on starlight, and that your life force is nothing but the energy of the sun! What religion could ever dream such a magnificent living connection between you and a star? That profound idea could almost be called "spiritual," if it weren't so entirely natural.

Nothing spiritual leaves the body when death arrives. Without breath and blood, cells gradually come to a halt over minutes and hours. Whatever is alive in the body that wasn't relying on the pumping blood, mostly foreign bacteria, continue to eat unimpeded by the immune system, and death brings its inevitable decay. But this only means that all that stored energy remaining in the body, all that captured starlight, gets converted by nonhuman organisms, and the endless cycle of vitality goes on. Starlight never rests, not here on earth, and nature wastes nothing.

How to Die Today

Death isn't what it used to be. Whenever humanity has gained new knowledge about how life works, revisions to the concept of death have always followed.

Prehistoric cultures didn't know about oxygen, but it was obvious that without enough blood, the "spirit" would not stay in the body, so checking for blood loss or a stopped heart was common sense. Traditional cultures the world over relied on this natural manner of checking for life: breathing and heartbeat. Nowadays, medicine calls this test the "cardiopulmonary" test for life: after a few minutes, complete loss of breath and heartbeat was a sufficient sign of death. And so it was—almost. Two rare exceptions were always possible. First, people stunned unconscious, heavily drugged, or severely injured can occasionally suffer from a complete cessation of breathing and heartbeat for a short time, before spontaneously coming back from that deathlike

stasis. That's why it was always wise to try to reawaken these people for a few minutes before giving up. They probably won't become zombies or join the undead, as far as science is concerned.

Nowadays modern medicine has a variety of techniques to restart bodily functions after many minutes. Second, unconscious people may fall into a state where they are barely breathing, and their heart is hardly beating, so that these organs haven't stopped entirely, but it is very hard to tell by just watching and listening to the body. Before the invention of the stethoscope to enhance the sounds of the lungs and heart so they can always be heard, no one realized how the body could fall into a near-death condition like that.

Still, such "false deaths" were extremely rare, and the criterion of death could remain pretty much what it always had been: without

breath and heartbeat for a time, death has occurred. This cardiopulmonary test remains the common standard for almost every society, and it
has only been supplemented by twentieth century medicine, and not
quite replaced. We now know that the brain is also vitally involved. But
the typical way someone gets declared dead nowadays anywhere in the
world is this: someone skilled at detecting breath and pulse won't be
able to detect yours for a few minutes, and that's how you die. If you
die in a hospital, hooked up to breath and pulse monitors, then those
monitors will silently testify to your death.

We are extremely interested in the brain nowadays for many good
reasons, but the odds that you will be declared dead because a wired-up
monitor of your brain activity settles down to a monotonous "flatline"
are extremely small. All the same, doctors rely on detecting brain activity in rare cases precisely because the brain can survive interruptions
to respiratory and circulatory systems, and the brain can even survive
interruptions to its own functioning. Death by definition should at
least be a permanent matter; so as long as life can resume after an unusual interruption, then death has not yet come. Biology recognizes
alternative modes of life, such as hibernation (a sleeping bear in winter), dormancy (like a seed), and metabolic stasis (like frog species that
can freeze solid during winter). Furthermore, the brain is an extremely
complex organ, so that deeper portions can continue to function even
though other outer portions like the cortex has largely or completely
died. That explains why people in permanent comas can continue to
live, since parts of the brain that sustain crucial life functions are still
doing their job, so that the heart, lungs, and other vital organs will
keep functioning without much medical help.

Only sophisticated technologies such as electroencephalography
(the EEG) can monitor which parts of the brain are still electrochemi-
cally active, indicating that brain functioning is still going on. After
even more extraordinary technologies were invented in the second half
of the twentieth century, permitting doctors to mechanically keep the
blood flowing in a body and the lungs processing air even though the
body would no longer do these things on its own, scientific medicine
had to think deeply about the role the brain has for life. After all, there

were now a few "breathing" and "pulsing" bodies lying in hospital beds that were "alive" according to the cardiopulmonary test, yet the EEG test said that no parts of the brain were alive.

The idea that who we really are as people depends on the brain is a relatively new notion. There were a handful of Greek investigators who strongly suspected that the brain wasn't as unimportant as generally supposed, but it took many centuries for the rest of the world to catch up. The Greek physician Hippocrates and the philosopher Plato agreed that the brain is the organ of the body most responsible for awareness and mentality. Aristotle preferred to prioritize the heart, as did many other Greek and non-Greek philosophers. Yet they were overruled by Galen's extensive studies of the human body, and his view that the brain controls the body's muscles through the network of nerves. Modern psychology now tracks the innumerable ways that the brain is necessary for life, directly and indirectly. Is the brain also necessary for death? Specifically, does medicine have to take into account the brain when determining death? Again, in almost all cases, the answer is no. However, in those rare cases when it is technologically feasible to keep the body alive with machines, we want to know whether we should bother. If there is reasonable hope that a coma may be temporary, for example, life-sustaining technologies can make good sense. However, if brain monitors reveal that the entire brain has completely died, without hope of revival, then the permanency of death become obvious.

Basing a medical definition of death on the best science cannot be avoided. Death isn't what it used to be, but that doesn't mean that we can't specify death in a pragmatic way for us now. Most countries today rely on the cardiopulmonary definition of death, supplemented by a brain-death test where necessary. The United States, for example, has used this two-part standard since the establishment of the Uniform Determination of Death Act of 1980. Most states have officially adopted this standard. It was designed and approved with the involvement of the American Medical Association and the American Bar Association, along with the Presidential Commission for the Study of Ethical Problems in Medicine and Biomedical and Behavioral Research. The key portion reads as follows:

An individual who has sustained either (1) irreversible cessation of circulatory and respiratory functions, or (2) irreversible cessation of all functions of the entire brain, including the brain stem, is dead.[5]

By law, or by customary practice, this two-part standard now supplies the test for death around the world. And that is how you shall be declared dead.

Notes

1. Alvord, Lori Arviso, "Navajo Spirituality: Native American Wisdom and Healing," in *Science, Religion, and Society: An Encyclopedia of History, Culture, and Controversy*, ed. Arri Eisen and Gary Laderman (Armonk, N.Y.: M. E. Sharpe, 2007), p. 669.

2. Ibid., p. 667–668.

3. Ibid., p. 668.

4. Grillo, Laura, "African Rituals," in *Wiley-Blackwell Companion to African Religions*, ed. Elias Kifon Bongmba (Malden, Mass.: Wiley-Blackwell, 2012), p. 124.

5. Uniform Determination of Death Act, Drafted by the National Conference of Commissioners on Uniform State Laws 1980, Approved by the American Medical Association, October 19, 1980, Approved by the American Bar Association, February 19, 1981. http://www.uniform-laws.org/shared/docs/determination%20of%20death/ udda80.pdf.

Further Reading

Boia, Lucian. *Forever Young: A Cultural History of Longevity from Antiquity to the Present*. London: Reaktion Books, 2004.

Bryant, Clifton D., and Dennis L Peck. *Encyclopedia of Death & the Human Experience*. Thousand Oaks, Cal.: Sage, 2009.

Cave, Stephen. *Immortality: The Quest to Live Forever and How It Drives Civilization*. New York: Random House, 2012.

Fridman, Eva, and Mariko Namba Walter, ed. *Shamanism: An Encyclopedia of World Beliefs, Practices, and Culture*. Santa Barbara, Cal.: ABC-CLIO, 2004.

Kastenbaum, Robert. *On Our Way: The Final Passage Through Life and Death*. Berkeley: University of California Press, 2004.

Keown, Damien, ed. *Buddhism and Abortion*. Honolulu: University of Hawaii Press, 1999.

Marcos, Sylvia, ed. *Women and Indigenous Religions*. Santa Barbara, Cal.: ABC-CLIO, 2010.

Miles, Stephen H. *The Hippocratic Oath and the Ethics of Medicine*. Oxford: Oxford University Press, 2005.

Palmer, David A., Glenn Shive, and Philip Wickeri, ed. *Chinese Religious Life*. Oxford: Oxford University Press, 2011.

Pettitt, Paul. *The Palaeolithic Origins of Human Burial*. London and New York: Routledge, 2013.

Riddle, John M. *Eve's Herbs: A History of Contraception and Abortion in the West*. Cambridge, Mass.: Harvard University Press, 1997.

Shushan, Gregory. *Conceptions of the Afterlife in Early Civilizations: Universalism, Constructivism and Near-Death Experience*. London and New York: Continuum, 2011.

Simmons, David S. *Modernizing Medicine in Zimbabwe: HIV/AIDS and Traditional Healers*. Nashville, Tenn.: Vanderbilt University Press, 2012.

Veatch, Robert M., ed. *Cross Cultural Perspectives in Medical Ethics*. Boston: Jones and Bartlett, 1989.

York, William H. *Health and Wellness in Antiquity Through the Middle Ages*. Santa Barbara, Cal.: ABC-CLIO, 2012.

Young, Katherine K., Harold Coward, and Julius Lipner. *Hindu Ethics: Purity, Abortion, and Euthanasia*. Albany: State University of New York Press, 1989.

8

SOCIETY AND POLITICS

Both religion and government evolved in response to the social problems presented by growing agrarian populations in the cradles of the world's early civilizations. It is a truism to suppose that government should at least defend the integrity of a country and promote the basic welfare of its people by enforcing some social order. What is religion's role in service to society? If religion and its history should be also studied in order to discover its contributions to social order, perhaps that study can also discern how religions have changed to remain relevant to maintaining social order. Both government and religion can ask, and answer, the basic question of how societies should be structured and maintained. From family life to national interests, religions have offered their wisdom about what the correct communal structure and social order ought to be. Some religions typically avoid politics, some religions are vague at best about government, and others seem designed to guide little else besides politics. Religions can be used to determine who shall have equal social standing and political citizenship—in a democracy, or perhaps an aristocracy, or even a monarchy.

If secularism is preferable, as many modern peoples think, what alternative grounds for human rights and political constitutions besides religion would be suitable?

The Social Order

The idea that the gods approve only specific kinds of social relationships and structures is one of the oldest cultural ideas that can be found across humanity. Previous chapters have recounted how religions endorse and enforce local cultural standards about male-female roles, marriage, the family, education, economic status, caste status, and the like. Later sections discuss religion and politics, and the overall theme remains the same: religions endorse and help enforce the dominant political hierarchy, whatever that happens to be. This close relationship, practically a merger, is pretty recognizable in most of the world's cultures.

Discussing religion's interactions with society or politics implies a conceptual distinction between them before one can look for relationships among them. One must be careful with how "real" one takes conceptual distinctions to be. There is an oft-heard complaint from historians and cultural anthropologists that religion must not be abstracted apart from the wider culture and social order, but instead regarded as thoroughly interfused with the other major aspects of life. This complaint is well-informed and quite just. The Western modernist ability to segregate religiosity and religious practices apart from other components of society, such as the economy or the government, simply reflects its own cultural path, one among many. Furthermore, no particular arrangement between religiosity and wider culture should be regarded as more "natural" or "normal" for humanity. Indeed, across humanity's history, most cultures blend religious teachings and practices with other important social activities. Those entanglements may be causing more harms than benefits, but making that judgment requires examining the evidence and deliberating with caution, rather than imposing a prejudiced elitism. Feelings of cultural superiority must be set aside. Most cultures have blended religion into most phases of life. Yet that prevalence doesn't justify religion's domination, any more than the West's global domination justifies bias against religion.

Judgments against religion must be based on religion's real effects on people's lives. This section looks at ways religion imposes its views on the civic lives of societies, and the next sections examine how religion has controlled politics.

Religions tend to support the prevailing social order, especially if that social order is stratified and aristocratic. Where does history record any dominant religion throwing its wholehearted support behind a populist movement trying to destratify social and economic power? Naturally, renegade portions of religions are found allied with attempted social revolutions. These reformist movements never end up in control of their parent religion, but must find their own way—usually as small cults, schismatic denominations, or new religions. The fate of new religious movements is usually bound up with the success of the social reform movement. A successful revolution empowers the new religion, while a failed revolution drives the new religious movement out into neighboring regions or straight into extinction. The biggest religions have been able to grow so large by relying on their enabling kingdoms and empires.

Hinduism's vast capacity for keeping lower classes in their place under aristocracies has never been surpassed. Buddhism, an offshoot of Hinduism, was received in a social matrix of lower class rebellion against the Hindu caste system, but it enjoyed immense growth under the reign of Buddhist king Ashoka (304–232 BCE). However, Hinduism soon reabsorbed most of Buddhism within India. Buddhisms beyond India survived, as the official religion of kingdoms or an optional religion for commoners under kingdoms. Confucianism was designed to support aristocratic and feudal societies, and admirably served kingdoms and empire dynasties across two millennia. Christianity was born from Jewish-Roman tensions, but whatever plans Jesus may have had were swept away, along with most Jewish-Christian churches, by the disastrous Roman-Jewish War of 66–73 CE. Only Paul's version of gentile Greek Christianity grew within the Roman Empire, surviving long enough to become the "official" religion of the empire three centuries later. The social and political causes for Islam's origins from the northern Arabian–Syrian region, where many cultures and religions were mixing together, are matters that remain obscure, but its expansion in the form of military empire is not.

Hinduism in India and Confucianism in China hardly ever supported populism, wealth equality, democracy, or even social justice. The recent phenomena of prodemocracy government in India and antidemocracy communism in China applied political ideas borrowed from Western civilization. Buddhism, Christianity, and Islam all made appeals to the low-born and underclasses during their early growth; all three ended up supporting feudal-military empires that kept the lower classes in their place. Anthropology finds that among religion's major functions is supporting the local social order, and this discovery is but a belated confirmation of the lessons of history.

Few religions overlook poverty; few religions, large or small, do anything to prevent poverty. Charity to the poor, and sympathy for the plight of the poor, has an essential place in religious thinking. Yet charity alone doesn't end poverty or grapple with the causes of poverty. Religions evidently feel at peace with the old saying, "the poor shall always be with us." The world still has plenty of very poor people today. Did religion help? In 1900, the world's population was 1.6 billion people. By 2000, it had reached 6 billion people. The twentieth century was truly a religious century—most of the larger religions enjoyed dramatic growth right along with the world's population, and there are four times more religious people now than in 1900. But did this explosion of religiosity make a difference to poverty? The twentieth century was a period of the most incredible growth of wealth the world had ever seen. In 1900, the estimated "gross world product," the value of everything produced by all countries added together, was 1.1 trillion dollars. In the year 2000, that figure was 41 trillion dollars. Where did all that wealth go? The world's richest countries in 1900 were still the world's richest countries in 2000, joined by a few oil-producing countries like Saudi Arabia. And the world's poorest regions one hundred years ago remain the poorest today, with few exceptions. Two-thirds of the world's population lived in poverty in 1900; about two-thirds of the world live in poverty now. The world's religions made no significant difference to the extent or distribution of poverty in the world as a whole.

The dramatic increase in wealth, and the way that this wealth remains concentrated among those already the wealthiest, simply reflect the way that capitalism, not religion, has been the far more powerful force on the planet in recent centuries. The world's religions are either too weak by comparison, or not concerned enough about poverty in the first place, to change the distribution of wealth or empower the powerless. Religions didn't lessen their fervent laments and pious prayers over poverty, not in the least. One can observe whether these good "works" actually made any difference. Religions did spend plenty of money on sending help to impoverished regions of the world, but most of that money was spent on missionary work and food distribution (or militarily diverted by corrupt governments). Relatively little has been invested into improving civil infrastructure or democratic conditions. In any case, aid to poor countries from religions haven't amounted to much when compared with the greatest sources of aid: secular governments such as America, England, France, Norway, and Germany.

244 · DISCOVERING OUR WORLD

Some fringe religious movements did lead activist efforts for the poor and dispossessed. The twentieth century will be long remembered for three powerful religious movements changing the real lives of the underclasses: Gandhi's protest against British rule over India, Martin Luther King Jr.'s civil rights movement, and the liberation theology movement inspired by Gustavo Gutiérrez, a Peruvian Catholic priest. All three movements were led by dynamic and charismatic individuals who used selected aspects of a major religion to inspire uprisings against social injustice. And all three movements had to deviate from religious orthodoxy and struggle against mainstream religion as much as the entrenched aristocracy.

Gandhi appealed to Hinduism's idea that all people share equally in a spiritual connection to the divine, regardless of their high or low birth on earth. Not an orthodox Hindu himself, as his parents were sympathetic to Jainism instead, Gandhi had the theological liberty to pick and choose from religious ideas, even Western ideas, such as his admiration for Ralph Waldo Emerson. His rebellion against British rule put his convictions about spiritual equality into action, organizing lower classes in both rural and urban areas in successful nonviolent "satyagraha" protests against colonialism. Hinduism couldn't be called an organized religion by any means, but Gandhi continually expressed his frustrations over the lack of support from influential Hindu leaders. Brahmin high priests and other leaders of great Hindu temples repeatedly rejected Gandhi's efforts to reduce prejudice against the lowest caste of untouchables, to promote intercaste marriage, and to end traditional practices such as child marriage. His collision with the Brahmins controlling the Vykom Temple illustrated Hindu orthodoxy's attitude. A satyagraha march including untouchables was blocked near Vykom during 1924–25 because the Brahmins refused to allow untouchables to use the roads. When public pressure eventually forced the Brahmins to relent, many Hindu temples followed that new example over the next decades, but thousands of temples did not. Brahmin Hindu orthodoxy was a continual obstacle to Gandhi's vision for full Indian democracy.

Was Gandhi's social movement religious? Surely it was religious in spirit as much as it was pragmatic in methods—but it cannot be said that his movement was mainstream Hinduism in action. His supreme ethical principle of the full equality of all persons in spirit transcended anything that orthodox Hinduism would have realized on earth, just as Buddhism and Jainism had already known two thousand years before Gandhi. His corollary ethical principles about modern society similarly transcend Hinduism to acquire a cosmopolitan status. Consider Gandhi's article for the magazine *Young India* (22 October 1925) which states his list of seven modern sins of society:

> Wealth without work
> Pleasure without conscience
> Science without humanity
> Knowledge without character
> Politics without principle
> Commerce without morality
> Worship without sacrifice.

Regardless of one's religious or nonreligious background, this list of sins should appeal to anyone as a source of genuine wisdom. Building a society to match these words of wisdom would require not one, but several dramatic social, economic, and political revolutions.

One such revolution arrived in America when marginalized Protestant churches, the black Baptist and Methodist churches in the South, organized protest movements against racial injustice. To say that Martin Luther King, Jr. did not have the support of mainstream Protestantism would be an understatement. His famous "Letter from Birmingham Jail" faulted prominent white clergy in the South for allowing their moderation to help perpetuate injustice and delay political change even longer. When the conservative voice says "Never" to change, and the so-called moderate voice replies, "Well, at least not now," then the combination of conservative and moderate voices only guarantees that little change will ever come, for this generation or the next. King's "prophetic" voice applied a Biblical perspective to contemporary events, viewing the struggle against racist oppression as the dynamic work of God in the world. Like Biblical prophets who compared God's law to their surrounding societies and judged that sin was in control, King judged America accordingly. He urged reconciliation with God's righteousness by establishing justice and political equality, along with rapid complete integration. King also appealed to another renegade Protestant's vision for human equality, designed by Enlightenment rationalist Thomas Jefferson two centuries earlier. This radical ethics of fundamental human equality, despite its political achievements, still hasn't become mainstream within Protestantism, exemplified by its internal divisions over women's rights to reproductive health, the "sin" of homosexuality, the fights over marriage equality, and the fact that churches are the most segregated places in America to this day.

Liberation theology couldn't gain mainstream Catholic status despite its own successes in South America. Despite the Roman Catholic Church's official disapproval and repeated censure by the Vatican, the liberation theology movement during the period of 1960s–90s drew serious political attention to poverty and social injustice. However, during the mid-1980s, the Vatican's doctrinal body of theologians under the direction of Cardinal Ratzinger, who later became Pope Benedict XVI, declared that nothing in the Bible or in Jesus's teachings are about empowering the poor, but only about giving charity to the poor. Liberation theology, according to the Vatican, was either Marxism or too similar to class-warfare socialism, so the Church must not get involved. Not that the Church historically has had difficulty allying with populist Fascism, of course. The Vatican signed treaties with both Italian dictator Benito Mussolini (1929 Lateran Treaty) and German dictator Adolf Hitler (1933 Concordat). The treaties brought immense wealth and investment opportunities to the Vatican, delivering a fortune now estimated at 15 to 20 billion dollars in financial holdings and properties.

Christianity has traditionally been far more comfortable with powerful and authoritarian forces. When your God is a heavenly King, your model for worldly politics couldn't be much different.

The Gods and Godly Kings

The idea that a king ruled with godly approval is already in place with the earliest signs of government in recorded history. As the earliest civilizations arose in Egypt, Sumeria, India, and China, and large political organizations were controlling large territories, there were kings and their gods. The settled tribes making up those earliest kingdoms already had their gods, we may assume, just as indigenous tribes down to recent times had their gods. Keeping these gods pleased with human affairs, and burdening tribal leadership with maintaining divine approval, is also far older than any civilization. With the rise of large kingdoms around 5,000 years ago in Egypt and Sumeria, and 4,000 years ago in Iran, India, and China, new relationships emerged between kings and gods. Prehistoric myths had credited gods with making the first people, and tribes everywhere assumed that their ancestors were descended from those first people. Tribal leaders many thousands of years ago may have additionally claimed special ancestral descent from gods, just as some have been recorded claiming in more recent times. The first kings of recorded history weren't content with a special ancestry going back to the gods. They wanted to be godlike themselves.

The first unified civilization in history was Egypt, becoming one nation around 3050 BCE. The first great king in recorded history was Egypt's first unifying king, who used the royal name Narmer. The first person in recorded history to claim to be a god was this same Narmer. When Egypt as a country begins with its northern and southern regions joined together, it is already using absolute monarchy as its government. That centralized control was firmly held together by a national religion worshipping the king as one of the gods. The god-king was humanity's first government.

This unusual precedence probably has much to do with Egypt's geography. Almost every settlement and city was near the Nile, a long river easily navigated most times of the year, so communication and travel throughout populated Egypt was not difficult. The major cities competed for internal power, of course, but once Egypt was politically unified and centrally controlled, it largely remained so, down to

the present day. Over those millennia, the idea of the god-king spread outwards wherever the Egyptian empire went, and even farther beyond its borders wherever its trade routes sent Egyptians journeying across Africa and western Asia. Other regions from North Africa, central Africa, the Near East, and Mesopotamia knew about the god-king. The idea of the god-king represented the ultimate power that a human could possess on this earth. Kings of other regions enviously desired such power, but few could achieve it.

The Mesopotamian kings, holding lordships over city-states scattered up and down the Tigris and Euphrates rivers, knew that their civilization was older than Egypt's, but they couldn't unify their lands so easily. Maintaining a dynasty over a single city, such as Ur, Kish, Nippur, Lagash, and Uruk, was difficult enough. Legends mention names of famous kings who are memorable for conquering their own city and perhaps a neighboring city for a while. Separating legends from history is nearly impossible, due to the scarcity of information. The earliest figures among these kings whose rule can be somewhat verified are En-me-barage-si of Kish (c.2700 BCE) and Gilgamesh of Uruk (c.2600 BCE). Only Gilgamesh was awarded divine status among all those kings, yet he remained partly human, and eventually died a mortal death. It was the southern region of Mesopotamia known as Sumer where writing, mathematics, astronomy, the calendar, the priesthood, vast temple-courts, and fortified city walls were all invented first. Egypt quickly imitated and developed every one of these innovations, far surpassing Sumerian achievements by the time Sumer was able to produce a great king, Sargon.

Sargon the Great (c.2300–2240 BCE) was the first king to politically unify all of Sumeria, along with most of northern Mesopotamia into Asia Minor, east into Syria, and west into present-day Iran. Unlike the kings of Egypt, Sargon and his heirs attempted to control an empire containing several ethnic groups. Also unlike the kings of Egypt, neither Sargon nor later great kings of Mesopotamia, including even the greatest rulers of the Assyrian empire of the north and the Babylonian empire of the south, proclaimed themselves to be gods. The only significant deification of kings in the region, during the city of Ur's third dynasty (c.2250–2000), is associated with Ur's decline and weakness, indicating an innovation based on desperation. This experiment was not imitated, and it abruptly disappears, suggesting an incompatibility with Mesopotamian culture. All the same, kings in general had little difficulty persuading the people that they enjoyed the favor of the gods. Upon their acquisition of power, they never failed to credit success to the supreme god of their home city, simultaneously proving that god's supremacy among all the gods and justifying their own supremacy over a kingdom. During their rule, these kings naturally attempted to control the religious priesthood, which was also the government bureaucracy, and many assumed high priest status themselves. But their religion did not automatically assign full divinity to whomever happened to be king. Kings made sure to drape themselves in symbols of their gods, great kings were often regarded as near-divine, and the greatest were fondly remembered as godlike for a while after their deaths. Now all but forgotten, eternal gods they were not.

Both Egypt and Mesopotamia agreed that from the very beginning of the cosmos, the supreme gods ruled over everything, and there remained a hierarchy of gods from the most important to the lesser gods. The major difference between Egyptian and Mesopotamian concepts of divine rulership is just as important. In Egypt, many gods helped to control how the world works in an orderly way, but they weren't responsible for maintaining social order in the human world, except for a single god, the Pharaoh. The Pharaoh was supposed to be just like the supreme gods who demand cosmic order and justice, because he was one of them. Egyptian political power ascended all the way up to a human god, who shared in the godly responsibility for keeping the world running correctly, by caring for human affairs. Everyone in Egypt served this human god, who didn't serve anyone else. In Mesopotamia, many gods were vitally interested in the political strength of societies, and they frequently intervened in human affairs by helping their own chosen rulers, the kings. Each king was supposed to obey the demands of the gods for cosmic order and justice, because he owed his position and power to them. Political power went all the way up to some god— for a while this god, then another god, depending on who was king— and that god delegated power to a human representative on earth who served that god. Everyone served this human king, who in turn served his god.

Next to the might of Egypt's Pharaohs, the most powerful rulers during the period of 1600–600 BCE were the sporadic unifiers of Mesopotamia, but neither Babylonia nor Assyria were able to consistently dominate the other. Assyria was able to dominate Egypt occasionally during this period, and then Nebuchadnezzar (c.634–562

BCE), the king of the Babylonian empire, was among the most powerful to challenge Egypt. However, not until Cyrus the Great (c.590–530 BCE) unified Persia (now Iran) and then conquered all of Mesopotamia and most of the Middle East, was there an emperor to rival the greatest Pharaohs of Egypt, such as Rameses II (c.1303–1213 BCE). Rameses the Great left many inscriptions on building and monuments across Egypt, declaring himself a god among the family of Egyptian gods. Cyrus the Great was probably worshipped by many of his peoples, but there were no temples for Cyrus, and his governing policy didn't include declaring himself a god. Instead, he relied on each part of his empire thinking that his right to rule was approved by whatever supreme god they already were familiar with. For example, Cyrus informed the conquered Babylonians that their own god Marduk had picked him out to take command. When Jews were permitted to return to Jerusalem, they interpreted this as a sign of Yahweh's work through Cyrus.

Cyrus's organization of his empire relied on a degree of regional autonomy under his governors, along with security for local priests who didn't cause trouble. This political latitude was interpreted as a measure of religious independence, since Cyrus had no god that he wanted to impose on the inhabitants of his empire. It may have never occurred to Cyrus to grant what we call religious liberty, since all he needed to worry about was obedience to his rule and the steady supply of resources to his governing capitals to maintain the empire. He simply left local custom and ethnic religion alone, since he didn't have to figure out how to change them. When he died, a single modest tomb was built in his homeland, and Plutarch recorded its epitaph:

> O man, whoever you are and wherever you come from, for I know you will come, I am Cyrus who won the Persians their empire. Do not therefore begrudge me this bit of earth that covers my bones.

This is not the grave inscription of a man who thought himself a god. His son Cambyses (c.559–530 BCE) did conquer Egypt for the Persian Empire in 525 BCE, promptly adding the title of Pharaoh to his list of rulerships.

God's Empire

Darius (550–486 BCE) was the next great Persian emperor, extending the empire to encompass all civilized areas from Macedonia and Egypt to the west, and all the way to the Indus river in the east. Only agricultural settlements and small towns, such as Athens and Rome, lay beyond the borders of the Persian Empire in Europe and Central Asia. Nothing but desert and shepherding nomads lay to the south in Saharan Africa and Arabia. To the east lay the Vedic-era kingdoms known collectively as the Mahā-Janapada, engaged in their own struggles for supremacy in northern India. Darius represented the heights of Persian civilization, now culturally and politically unified under Zoroastrianism. He regarded himself as appointed by the supreme Zoroastrian god Ahura Mazda. Yet he did not attempt to impose this religion across his empire. Instead, he following the liberal policies of Cyrus. However, Darius was happy to accept worship as a god in Egypt, where he held the Pharaoh's throne, and he perpetuated the identification of the kingship with the supreme religious authority.

Darius's successor Xerxes (519–465 BCE) and subsequent Persian emperors settled into a three-fold policy: accept worship as a god, or at least having divine ancestry, where regional custom preferred it; let local religious practices prevail so long as the supremacy of the emperor is acknowledged; and increasingly identify one religion, Zoroastrianism, as the ultimate ruling religion with the emperor as its high priest possessing some divinely bestowed powers. Zoroastrianism was the ideal kind of religion for the policies of this Persian Empire. Instead of a polytheism involving many competing gods in an unstable hierarchy, Zoroastrianism had only two supreme gods, one good god named Ahura Mazda, the emperor's god, and one evil god who caused disruption. Local gods were either loyal to Ahura Mazda to signal a region's obedience to the emperor, or loyal to the evil god to signal a region's rebellion. Zoroastrianism is not a true monotheism, since it has a supreme god but accepts the existence of other subordinate gods. A term useful for labeling this kind of religion is hierarchical henotheism—all gods are ultimately dependent on one supreme god.

Xerxes, like Darius, was more concerned about economic and military challenges to the empire than anything religious. Their antipathy against the rising Greek power in the Aegean was motivated by a worry that the Greeks would spread into Asia, just as they were colonizing lands around the Mediterranean Sea. Failing to conquer Greece, they only aroused a wakening giant to action. The Macedonian king Alexander the Great unified Greece and then, armed with impressive Greek military technologies and his own genius, conquered the entire Persian Empire by 325 BCE, reaching as far east as the Ganges River in India. Proud of his Greek heritage and culture, Alexander speedily built many new Greek cities and added a layer of Greek culture to conquered cities from Egypt and Syria all the way to India. However, the cultures he conquered also influenced his rule. Upon capturing Egypt, he was pronounced divine Pharaoh and the son of the supreme Egyptian god Amun. In Babylon and Persia he was treated the way the Persian emperors were treated, as the divinely appointed religious ruler, most worthy of submission and worship. Even back home, some Greek cities offered to regard Alexander as a semidivine son of Zeus.

Stories that Alexander encouraged his own deification are probably exaggerated by ancient historians. Contemporary historical judgment can only say that Alexander couldn't have really thought of himself as a god, but he did allow the policies of the Persian Empire to continue. If some regions wanted him to be a god or be godlike, helping to keep the empire unified, Alexander didn't object. Personally, Alexander remained Aristotle's student. As Aristotle taught in his philosophy, there is one supreme hidden god, responsible for nature's basic energies, and humanity was made by nature, not Olympic gods. Nor did Alexander violate Aristotle's political views—across his empire, he preferred the model of the self-governing *polis*, the Greek city-state, and an empire bound together in strong federation and intercultural integration. However, Alexander's death at a young age prevented him from shaping his empire according to his vision.

The civilized world from Italy and North Africa all the way to India now had a single idea about what a real empire looks like, and what holds an empire together: the single monarch beloved by the supreme god and wielding godlike powers.

The Romans couldn't fail to notice. The idea that kings were close to gods wasn't new to them. Their origins in tiny towns were mythologized into grand narratives, having themes and plots typical to Bronze Age cultures. Their legends about Rome's founding, the ancestry of aristocratic families, and a heritage of early divine kings were religious narratives pointing to important gods behind it all. Ancient Roman religion, like Greek religion, was also a kind of hierarchical henotheism, since many gods all descended from, or at least obeyed, one supreme god such as Jupiter in Rome or Zeus in Greece. However, Rome invented its own blended form of aristocracy and popular representation in a republican form of government to replace kingship by 500 BCE. Essentially, instead of continually fighting over which aristocratic family will hold the monarchy, they agreed to share power, which turned out to be a far more effective way to enlarge and control the Roman territory. If the growing Roman nation had halted before conquering vast territories from Spain up to France and east to Turkey and Egypt, it might have remained a republic for a long time.

By the time Rome realized that it had become an empire, with a vast military institution effectively running most of it, the problems sustaining that immense military had become overwhelming problems for the government. When the military's problems become the government's principle problems, dramatic change becomes inevitable, since a vast military system is too hierarchical and too powerful for anything besides monarchy. By the time Julius Caesar became a military commander in 58 BCE, Rome had been already been convulsed by civil wars fought by military generals demanding supreme power back in Rome. Caesar eventually demanded, and received, status as lifetime dictator over Rome in 44 BCE, but his prompt assassination leaves only speculation as to whether his monarchy would have evolved towards deification.

I'm not a god, but I could be...

There is no need to speculate what Caesar's successor Octavian would do. Upon receiving the title of Augustus in 27 BCE, he guided all parts of the empire towards respecting him as divinely approved and destined to join the gods upon death. Because the next emperor would be chosen from the emperor's sons (natural-born or adopted), this system of deification upon death guaranteed that each current emperor was the son of a god, somewhat divine in nature already, and fated to join the gods. (If an upstart or usurper became emperor, a prompt deification ceremony for his ancestor was conducted.)

This official political religion blended together lessons learned from all parts of the empire. For centuries, Roman governors over tribalistic and less-civilized peoples in the provinces had gotten used to being the inevitable object of cult worship and deification. Governors over civilized regions such as Egypt, Syria, and Babylonia were treated to customary semideification there as well. A provincial ruler over Egypt knew what it was like to be popularly regarded as a divine Pharaoh, just as the previous Greek Ptolemaic rulers had encouraged, down to Cleopatra herself. However, the Romans, having followed the somewhat naturalistic outlook of the Greeks, couldn't quite regard a living emperor as a god, and direct worship and sacrifice before an emperor was considered excessive. However, the emperor held all major religious offices, and the Roman religion and government were effectively fused together. Cults and temples to the emperor's family and Roman gods, which never failed to include a statue of the current emperor, were placed in every corner of the empire, and no locality could fail to show public respect. By the time of his death in 14 CE, Augustus had acquired the greatest political and religious status on the planet since the days of Pharaoh Rameses II, thirteen hundred years earlier.

Like the Persians, the Romans only insisted on the emperor cult's superiority to local religions, and Rome never regarded local religions as something to be eradicated, unless it rejected official worship practices. It was Christianity's strict monotheism, not its humble provincial origins or its spread into Roman cities, which set it on a course of conflict with Rome. Where Jesus is depicted as saying, "Render unto Caesar the things which are Caesar's, and unto God the things that are God's" in Matthew 22:21, a second century attempt to portray the young religion as harmonious with Rome was under way. Yet many Christians did renounce respect for Roman temples, and antagonism was inevitable. All the same, Christianity didn't reject the power that the emperor's throne represented. By the time emperor Gratian declared Christianity as the official religion of the Roman Empire and began punishing apostasy from Christianity around 380 CE, the emperor's status as the supreme religious figure holding semidivine power had long been standard practice. Christianity now had to wield supreme political power, and Christians justified that power by judging that their God's will was responsible. Because this one true god was ultimately in power and demanded exclusive worship, Christian emperors began to abandon the Persian and Roman toleration of local religions. Christian missionaries were able to freely travel across the provinces and eventually into the "barbaric" lands beyond. The Christian emperors stopped promoting their own divinity, but they were distracted by the division of the empire into the Western Roman Empire and the Byzantine Empire after Constantine's death.

In both the east and the west, political and religious supremacy were separated. In the west, the last Christian emperors did not pretend to be the supreme religious authorities, as the Roman papacy took command of the Christian Church. In the east, the Christian emperors remained influential over the Patriarchs, although no single "pope" was recognized there, throughout the course of the Byzantine Empire until its collapse to Islam in 1453. Islam's conquering of vast territories from Spain to Iran didn't result in a return to the ancient Persian or Roman policies of erecting divinity cults, in this case around Muslim caliphs and dynastic rulers, since Islam's strict rejection of divinity for humanity held firm. In Christian realms, east and west, no king or emperor claimed semidivine or divine status, and no pope or patriarch claimed to hold supreme political power. Instead, these two centers of power began to forge alliances, which were as likely to cause conflict as cooperation.

Monarchs needed the appearance of divine approval and Church support, so papal approval was quite useful. For their part, many popes were tempted to ally with strong monarchs to protect their own power, and preserve the appearance that they alone could mediate between god and society. During the chaos of the "Dark Ages," a weak pope allied with king Charlemagne, who then took the title of Augustus, Roman Emperor, in 800. The concept of a Christian-Roman empire over Europe gained strength when Pope John XII crowned Otto I as Holy Roman Emperor in 962, an institution that lasted until 1806. The only major European monarchy to unify supreme political and religious power was England's, after Henry VIII declared himself to be supreme head of the Church of England in 1534. French monarchs, especially the most powerful Louis XIV (1638–1715) came the closest to absolute monarchical power over church and state. The rationalism of the Enlightenment infused France's secular revolution in 1789, and even the return of kings to France didn't revive a cult for kings. When Napoleon of France was ready to be crowned emperor in 1804, the pope was present at the coronation preparing to place a crown upon Napoleon, but Napoleon simply took the crown and set it on his own head, for significant symbolism.

Napoleon was defeated at Waterloo in 1815 by the English army of King George III and the Prussian army of Frederick William III, culminating the first great European war to be fought in which none of the embattled kings claimed divine right to their thrones. All three countries continued to have kings for a time, but democracy was expanding, and by the start of the twentieth century, democracy (for men at least) was firmly in place in all three countries. The only significant ruler over a European territory who continued to claim that God alone gave him the right to rule was the Pope of the Vatican.

Heavenly Rulers
Alexander the Great's invasion into India was stopped when his armies refused to advance farther in 326 BCE, fearing the powerful Nanda Empire. After Alexander's death, his generals couldn't hold the hard-won territories for long. The Nanda Empire, the first to unify most of northern India, was soon conquered from within by something Alexander couldn't have predicted: another young military genius. A legend that Chandragupta Maurya (c.340–290 BCE) actually met Alexander as a youth captures the sense of astonishment at two conquerors occupying the world's stage one right after the other. Chandragupta unified the entire Indian subcontinent, and established his Maurya dynasty. No "pope" of Hinduism, if one were even possible, assisted Chandragupta—his chief advisor was the administrative and political expert, Kautilya, whose treatise *Arthashastra* stands among the world's great works about political practice and government administration.

That dynasty peaked in territorial extent and power under his grandson, Emperor Ashoka (c.300–232 BCE). Ashoka converted to Buddhism during his reign, because of his remorse at the immense loss of life during a war campaign. Ashoka recounts his conversion and refers to himself as "Beloved of the Gods" in his Edicts, which can be read today as inscriptions carved into large pillars and rock formations distributed around India. Ashoka urges conformity to the "Dhamma"—a term in Ashoka's language of Prakrit which approximated what Dharma meant for Hindus in Sanskrit, referring to the highest religious duties in conformity with the cosmic divine order. Ashoka elevated what had been religious duties for personal "salvation" into a community ethics governing the peoples of his territories for their benefit. While Ashoka probably did not become a Buddhist monk himself, he was a practicing Buddhist for the rest of his life, and he looked forward to an afterlife, although his edicts don't describe what that afterlife might be like. His vagueness is consistent with his official toleration towards religions. Ashoka used his throne to preach Buddhism's basic personal and community ethics, but he also commanded that all religions should be understood and respected. His restraint from imposing Buddhist practice on his empire was based on practical necessity, since Hinduism remained well-entrenched in India. The simple Buddhist ethics that Ashoka promoted, revolving around respect for animal and human life, a life of moderation, and obeying family and civic duties, was not much different from Hindu ethics, with the notable exception that Ashoka tried to restrict the animal sacrifices common to Hindu rituals during that era.

There is no evidence that Ashoka deviated from enforcing the rigid social order that Hinduism provided with its caste system. This system of divinely commanded *varna* in the ancient Hindu Vedic scriptures specified four *varna* (castes): the Brahmins (priests), the Kshatriyas (the military rulers), the Vaishya (land owners, merchants, craftsmen), and the Shudras (all other laborers). Besides the four castes, the Untouchables, mostly made up of the unassimilated indigenous peoples conquered by Vedic-worshipping tribes, were deemed to be beyond the *Varna* system with a status barely higher than animals. The ancient system was a form of racist supremacy in which lighter-skinned people were entrenched in the upper castes in order to parasitically live off the near-slave labor of lower classes. In accordance with *Varna*, Brahmins were treated with great respect by rulers, and the holy rituals conducted by Brahmins certified divine approval upon rulers so that the masses would obey. Not coincidentally, the immense scale and quantity of animal sacrifices and other forms of material goods surrendered in order to keep the gods satisfied simply went straight to feed the ruling classes' appetite, so that popular Hinduism justified economic extortion as well as political sovereignty.

Kings were not regarded as gods according to this religious-political system, but the most powerful kings and emperors were viewed by the masses as possessing semidivine status and godly powers so long as they upheld the Hindu Dharma. There is little evidence that the priests or kings themselves regarded any humans as gods. Priests did not, as a rule, make sacrifices to kings, and even kings had to be concerned for their own karmic reincarnation instead of getting a divine exemption from the endless cycle of death and rebirth. Hinduism provided for a truly godly kind of being who was capable of making appearances on earth in human form: the Avatars. Descriptions of these Avatars, such as Krishna and Rama in the epic works *Mahabharata* and *Ramayana* (containing stories as old as 800-600 BCE), supplemented Vedic theology with explanations for appearances of saintly Hindu figures across many centuries. The Brahmin caste had the Vedas, but they lacked a unified method for maintaining doctrinal purity across an entire subcontinent. Instead, they resorted to another version of hierarchical henotheism, by adopting a permissive attitude towards all local gods, so that Hinduism encompassed thousands of gods and innumerable local saintly figures. Hinduism only insisted that all of them are earthly manifestations of the supreme pantheon of Hindu gods, and ultimately of the mysterious transcendent god Brahman, according to the later *Upanishads*. Nowhere did Hinduism have any incentive to impose a single small pantheon or lone exclusive god upon all of India, since the priestly-military alliance only required that local inhabitants accept their local king as legitimate according to the local gods they already worshipped.

Hinduism didn't develop into the sort of religion expecting one supreme god to appoint one supreme ruler over India. Bound together in a pact of mutual support, sanctified by Hinduism's holiest *Vedas* and enshrined in the Brahmin rituals, the aristocratic alliance between the priesthood and the military was rarely challenged throughout India's long history of dynastic kingships down to British colonial times.

While the Ptolemaic pharaohs in Egypt waited to join their fellow god Amun-Re, Xerxes ruled the Persian Empire under Zoroastrianism, and the Mahā-Janapada kingdoms sought Hindu gods' approval, the Warring States period in China was starting as the imperial Zhou dynasty continued to disintegrate. Master Confucius (c.551–479 BCE) observed with disgust that it seemed his culture was falling apart. He knew exactly where to place the blame: rulers were changing, or forgetting, the traditional religious rites and ceremonies that bound society together. The primary collection of his sayings, the *Analects*, repeatedly urges rulers to return to the traditional ways and the traditional social order which had held society together.

> *Analects* 3.9. The Master said, "As for the rites of the Xia Dynasty, I can speak of them, but there is little remaining in the state of Qii to document them. As for the rites of the Shang Dynasty, I can speak of them, but there is little remaining in the state of Song to document them. This is because there is not much in the way of culture or moral worthies left in either state. If there were something there, then I would be able to document them."[1]

Confucius accuses Zhou states during his own time of largely abandoning the traditional rites dating from before the Zhou, from the two earlier dynasties of China: the Xia Dynasty dating before 1600 BCE and the Shang dynasty which had ended by 1000 BCE. Besides restoring respect for these old ceremonial rituals, the Confucianism explained in the *Analects* prescribes the correct virtues and duties for all members and classes of society, especially the rulers. Community participation in all the customary rituals is the only way to restore obedience to the traditional morality which those rituals teach.

Confucius occasionally urges conformity to ways of Tian, the heavenly skies. Yet Confucius expressed very few ideas of his own about this heaven. He was confident, as the Chinese had always believed, that great ancient kings, the "sage kings," continued on as this divine Tian, serving as spiritual guides and models of exemplary virtue and knowledge. These sage kings were largely responsible for organizing nature's ways and bestowing skills and technologies upon humans to make them different from animals. The cosmological order is therefore the model for customary society. The kind of heavenly immortality enjoyed by these sage kings was only available to kings of great wisdom and power. Only after bestowing a legendary status upon him long

I know, but I'm not telling.

after his death, would anyone regard a king as heavenly. No actual king attempted to claim divine or heavenly status—however, having an ancestor among Tian was crucial for maintaining at least an appearance of legitimacy.

Imitating Confucius, later writers agreed that following the will of heaven, which meant imitating the exemplary sage kings by maintaining social order and enjoying successful reigns, is the top priority of any king or emperor. In that way, a ruler maintains cosmic order for his people on earth, entitling him to be a "Son of Heaven." Every ruler must pass the same practical test: demonstrate proper authority as the benefactor of the people in order to receive heavenly approval.

Borrowing from older mythological ideas common to the period of the Zhou dynasty, inaugurated around 1000 BCE, the *Analects* expect that rulers should follow the fine examples of these semimythical kings. Successful rulers enjoy their approval. Warnings from heaven, such as natural disasters, famines, and civil war, signal heavenly disapproval and justify the overthrow and replacement of poor rulers. The Zhou dynasty had justified its overthrow of the previous dynasty by claiming that the "mandate of heaven" had bestowed legitimacy upon it, twisting much older Chinese traditions of obedient ancestor worship into political propaganda. But the dynastic kings were no gods themselves, not during this mortal life. And the bureaucrats which Confucius hoped to train for careers advising rulers were not intended to make kings into gods fit for heaven, but rather into wise patriarchs serving their peoples. The Han dynasty (202 BCE–220 CE) solidified the administrative model for later dynasties. That model required an emperor using a vast imperial court of bureaucrats, capable of ruling over a feudal empire consisting of agricultural fiefdoms in turn led by hereditary lords from aristocratic families.

Chinese religion remained fragmented and disorganized. Extremely polytheistic going back to prehistoric times, popular religion was full of nature spirits, malevolent ghosts and demons, and an eternal heavenly realm for a vaguely supreme god, called Shang-Di, along with the sage kings. The Chinese never even attempted hierarchical henotheism. Heavenly gods were not regarded as interested in having ordinary people join them after death, nor were the sage kings interested in trying to command and control kingdoms, or in communicating with rulers about ordering society. The notion that some ancient sage king would want to still be a "king" of an empire or use a human king as a tool to stay in spiritual power would have been a ridiculous notion to the Chinese. The warlords and emperors never had any incentive to impose worship of just one sage king or one god upon their peoples. Without any direct commands arriving from heaven, the only option was to attempt to "read" signs of cosmic approval or disapproval by watching natural phenomena and conducting divination rituals to predict the future. The popular religion of Daoism evolved from those same roots, providing a sense of cosmic order revolving around one eternal power called the Dao, and offering traditional wisdom for maintaining personal health and social tradition, but nothing in Daoism would serve kings trying to stay in power. Similarly, when Buddhism arrived from India the common people absorbed some of its lessons for understanding life and death, and possible reincarnation. Yet no Buddhist kings arose in China to attempt to make China uniformly Buddhist.

A courtly religion revolving around ceremonial sacrifices and the worship of heavenly figures served to keep the Chinese aristocracy in power. No powerful priesthood emerged in China to share power with the kings. The courtly priests responsible for maintaining the rituals never could become an aristocratic class in their own right. They had no ability to deliver divine powers or instructions down to earth, nor did they have any way to help kings obtain divine status. The lone exception to human kingship, the proclamation by the first Qin king (starting in 221 BCE) to be divine, was more about bypassing older aristocratic lineages, and hence Tian itself, in order to consolidate absolute power. This dramatic revolution was not taken seriously, ending quickly after the first king.

By the time of the Han dynasty, the courtly priests were gradually replaced by an educated class of bureaucratic officials devoted to practical administration in philosophical accord with traditional ways. To this day, religiosity in China has remained politically weak, since nothing in Chinese religion could assist any ruler with obtaining some godlike status or special communication with gods. Nevertheless, the feudal system under dynastic kings was durable enough to last until the twentieth century.

Religion and Social Stability

Across ancient and medieval civilizations, religions tended to ally themselves with the dominant warlords and powerful kingdoms. Religious leadership always has strong incentives to collaborate with aristocracies and rulers, to either acquire aristocratic status for themselves, or to at least enjoy government toleration of their parasitic practices upon credulous populations.

Where strong political authority expends its governing power for rigidly enforcing religious belief among the people, prosperous kingdoms and nations typically have very high religious conformity. There are past and present-day examples. Medieval Europe is one familiar example, in which every stable and wealthy country from Sweden to Italy had virtually uniform Catholic religiosity. Powerful governments in alliance with a powerful Catholic Church made it almost impossible to be nonreligious in those societies. The culture was so thoroughly religious in almost every respect that bad government couldn't harm religiosity, either. Through every sort of economic hardship, social collapse, and natural disaster, despite famine or flood, religiosity rarely was questioned in a culture of intense religious conformity. Good government or bad, society remained very religious, since the prevailing conviction remained that only a unified religious culture makes social order and government even possible.

During most of the course of civilization on earth, four main types of religion-politics relationships are prominent. One type is the Egyptian/Medieval Christian/Classical Islam option of pairing absolute monarchy with monotheistic religion, so that everyone must observe the same religion in order to display full obedience to the ruler. (Tolerant Muslim Caliphate empires accepted Christianity and Judaism as fellow "Abrahamic" religions.) A second type is the Persian/Hindu option of pairing a federated empire with a polytheistic religion, so that people can worship their local gods so long as they obey regional rulers subordinate to an emperor. A third type is the Roman/Chinese option pairing a federated empire with a courtly religion and toleration of popular religions, so that the people can have their own religiosity so long as they participate in the civic rituals as well. A fourth type is the modern Western option of pairing a strong republic with toleration for all religions, so that equal citizens can respect good government, and replace poor government, regardless of which religion they freely practice.

Only the emergence of this fourth option of popular government with religious liberty permits popular religiosity to be completely independent from government power, at least in theory. That means that in countries where good government maintains security and grows prosperity, such government can enjoy popular support even though it doesn't promote one religion, or any religion. In fact, religiosity can change dramatically over time, as some religions grow while other decline or new religions arrive, without affecting the ability of a government to do its own secular tasks. Europe had to transition from a medieval model to a modern model by passing through a very difficult period of terrible religious warfare. Protestant Christianity seemed to look like a grave threat to Catholic monarchies, precisely because the medieval Christian model presumed that all good citizens had to belong to the same Church.

We may compare the eruption of Protestantism in Europe in the sixteenth century with the spread of Buddhism into China by the fourth century. Buddhism was sporadically oppressed by a few kings intolerant of foreign religion, but nothing in Buddhism provoked Chinese concern that it would overthrow governments and use kingdoms to propagate itself. Buddhism's divine being(s) are uninterested in ruling kingdoms and Buddhist monks have no divine commands to relay to rulers. (Medieval Buddhist kingdoms in the Burmese, Cambodian, and Thailand regions imitated the Hindu model of politicized religion, since Buddhism had no innate political theory.) As for Christianity in Europe, its strict monotheism and aristocratic priesthood made it a much more potent factor in politics. After the Protestant schism, it took three centuries of religious war in Europe until many kingdoms restabilized under the novel theory which took root that government doesn't have to impose the same religion on all citizens in order to perform its functions properly. Governments can defend the home-land, deliver justice, promote opportunity for all, and so forth, without obeying or helping any one religion over its alternatives.

Any theory about a historical relationship between religion's effects on culture and stable social order has to take into account the differ-ent religion-government models. Where governments believe that they must regulate religion in order to survive and function well, and keep the aristocratic class in power, these governments usually work to en-sure high levels of religious participation in a single religion. The best pairing in that effort is a religion that is already somewhat monotheistic like Christianity, or at least hierarchically henotheistic like Hinduism. Where governments do not suppose that regulating their population's religions is necessary to function, such as the Roman and Chinese em-pires, and contemporary Europe and America, the religiosity in those populations can be very diverse and irregular according to local inter-ests and conditions. Periods of intense religious pluralism, interchange of religious ideas, and religious creativity over the past 3,000 years have flourished during phases when a large empire is neglecting or unin-terested in popular religiosity—and especially when a large empire is disintegrating or merging with a neighboring civilization. Much of the religious/philosophical creativity associated with the "Axial Age" of 800–200 BCE correlates with empire disintegrations, civil wars, and cultural openness from Greece and Babylonia to India and China. The eastern Mediterranean region during 300–600 CE is a later exam-ple, as the Roman Empire collapsed—mysticisms abounded, Persian Mithraism and Gnosticism flourished, Christianity coalesced into its Catholic form, and the ingredients of Islam were coming together.

During the past four hundred years, the single greatest impact on civilizations around the world has been the immense technological power and economic might of European empires such as Spain, France, England, and Germany, followed by the American financial empire. This power has elevated material prosperity—bringing associated ben-efits like social welfare, personal health, and longer lifespans—for these countries and their populations. A list of the most prosperous nations as a whole in the year 1700 would have to include England, France, Italy, Germany, Russia, India, China, and Japan. Because the populations of these countries varied enormously in size back then, from around 6 million people in England up to as many as 140 million in India, the amount of wealth per individual person varied greatly. The "average" individual in England or Italy had far more wealth and prosperity than most people living in India or China, and by 1800 the disparity was dramatically accelerating. By 1900, most northern European countries, along with former British colonies such as the United States, Canada, and Australia, had increasingly secular and democratic governments, which maintained social order and stability without enforcing religious conformity.

These countries kept accelerating in material prosperity and continue to be among the wealthiest countries in the world today. These are also the countries with the lowest measured levels of religiosity in the world, joined by Japan, China, and Vietnam. These three Asian countries do not have a history of governments reinforcing one religion upon their populations, so their levels of devout religiosity are still quite low, and there is little correlation between prosperity and religiosity. For example, Japan has four times the per capita income as China but a similarly low religiosity. Even the United States has fallen to the level of 60% of people reporting that they consider themselves to be "a religious person," according to the 2012 WIN-Gallup International Religiosity and Atheism Index. Still, that combination of high wealth and relatively high religiosity in the United States stands out, as do the other two examples of countries having a per capita income higher than $20,000 with levels of religiosity higher than 60%: Saudi Arabia and Italy. These three countries could be called "outliers" or even "counterevidence" to the theory that prosperity implies lower religiosity, but none of them really disprove anything. Saudi Arabia is extremely religious, but it is rich only because of its fortunate supplies of oil, and not because of anything its monarchy has done for the people except distribute a small amount of its wealth, keeping most for its aristocracy. Italy is the home of the Roman Catholic Church so it has been fairly religious, although its religiosity is falling too, not far behind other predominately Catholic countries like Spain, France, and Ireland, which fell below that 60% level years ago. In the United States, religiosity has been falling for two decades and will reach the levels of more secular countries such as Canada and Germany within another generation, if trends continue.

What about correlations between low prosperity and religiosity? According to that same 2012 survey, the ten most religious countries in the survey had per capita incomes of less than $14,100. Other international polling has obtained similar results. A Gallup survey in 2009 revealed something unusual about the poorest countries, those with less than $2,000 per capita income: all of them had extremely high levels of religiosity. Furthermore, when additional measures of social and personal well-being are taken into account, the pattern continues. Where countries have high levels of crime and murder, wide income gaps between the poor and rich, high rates of child mortality, lower life expectancies, teenage pregnancies, large family sizes, and similar signs of bad social health, these countries have the highest levels of religiosity in the world. Furthermore, these countries typically suffer from weak, unjust, or corrupt governments unable to guarantee justice and prevent aristocracies from robbing the country.

Using the label of "social fragility" to point to the combination of low prosperity, poor social well-being, and social injustice, then data from around the world displays how social fragility, not social order, is found wherever high religiosity is found. When a country like the United States stands out among wealthy countries for its high religiosity, social fragility supplies an explanation, since the United States also stands out among those countries for its far higher crime and murder rates, its wider economic gaps, its lack of accessible health care and affordable education, and so on. Its greater social fragility among similarly wealthy countries, resulting in higher levels of religiosity, follows the general pattern seen around the world. This strong correlation between social fragility and high religiosity has no serious counterexamples. It might appear that China is a counterexample, since it has a low per capita income and low religiosity. However, the Chinese religion-government model would also indicate that popular religiosity would not rise there simply because of poverty and some instability, since Chinese people typically don't expect religiosity to magically bring wealth, manipulate the government, compensate for social injustice, or deliver heavenly rewards in an afterlife.

One further point can be made here. Some scholars have pointed to the United States' secular government to explain its high religiosity, on the notion that where government is tolerant towards any religion, its people will more easily be religious in a manner of their choice. This economic supply-demand model of religious choice can't get much confirmation from looking around the world, and it cannot explain religious demand in the first place. Most northern European countries have had religious freedom for almost as long, or even longer, than the United States, yet their religiosity declined sharply decades ago as their secular governments greatly reduced social fragility. Where there is higher social fragility there is higher religiosity, and the need to somehow deal with social fragility is a better explanation for interest in religion than just the freedom to be religious. Furthermore, in nations and empires where there has been very little religious freedom, very high religiosity has also been observed, so mere freedom does not create demand or religiosity.

Although many civilizations in world history can be interpreted as confirmations of the theory that religion contributes to social order, it is also the case that religions can grow and flourish during periods of political instability and social fragility. What is the best way to view the relationship between religion and social order? Evidently there are multiple types of religions, diverse ways that religions and governments can relate, and even the same religion can flexibly adapt to different political conditions. For example, monotheistic Christianity can flourish during empire disintegration (the fading of the Roman Empire), extreme poverty and chaos (the Dark Ages), the political rigidity of absolute monarchy (Medieval era), international warfare and religious wars (1500s–1700s), and modern conditions of secular democracy. The flexibility of a religion could be interpreted favorably as a sign that religion is still contributing to social order during times of disorder, as a helpful remedy leading on towards better stability eventually. This theory in effect says, "Without religion, it would have been much worse!"

However, the global evidence shows that many religions have flourished precisely where terrible social conditions have lasted for centuries without much improvement. It is hard to see how things could have gotten worse, with or without religion, during such terrible eras—and religion was also perpetuating religious wars, making civil wars more hateful, keeping people ignorant about medicine and science, denying education for women and children, encouraging large family sizes, and so forth. Then the inescapable judgment logically follows: these religions are actually taking advantage of low prosperity and high social fragility. In fact, these religions are actively working to sustain the very conditions that make religiosity seem so useful to people so deprived. It is not a coincidence that religions which have settled deeply into cultures of poverty and instability typically worship gods demanding faith over education and submission instead of empowerment.

The remedy for these parasitic religions seems obvious: reduce social fragility by strengthening secular government. Ideally, all governments should become strong secular and democratic governments that elevate wide prosperity (not just for the rich), deliver legal, economic, and social justice, and generally keep social fragility as low as possible. Religions compatible with toleration, democracy, and education can still operate—call them "domesticated" religions—but parasitic religions cannot survive under those ideal conditions. The parasitic nature of religion never goes away, however, so constant vigilance is necessary. Even in a prosperous democracy, when social fragility increases, then religion can turn parasitic once again. Wherever people feel afraid for personal safety or threatened by hostile foreigners, or people feel insecure about their income, or people are losing access to education and health care, then parasitic religion emerges to offer its consolations and promises—available to those willing to donate generously, of course.

There is no single pattern to the relationship between parasitic religion and politics. Parasitic religion in itself is neither "conservative" nor "liberal." Fringe parasitic churches can appeal to traditional values and customary ways, or they might be appealing because they cultishly offer radical or communal forms of living. An evangelical fundamentalist church, a "hippie" new age cult, and a remote monastery appear to have little in common, but they do supply an escape from the frustrations of mainstream society. However, there is a natural fit between conservative parasitic religion and political conservatism in general. When they find each other to combine forces, not only does conservatism try to politically preserve traditional privileges and old social orders against the tide of democratic progressivism, conservatism also tries to religiously obstruct secular government from protecting the middle and lower classes.

Social fragility is a serious problem in itself. Add religion, and matters usually get worse. Religion makes no investment in local infrastructure, nor does it infuse economic opportunity into an impoverished region. Religion primarily takes money from poor people to erect religious buildings and pay salaries to clergy busily harvesting more souls along with their money. From an economic standpoint, placing a heavy tax on poor people, or setting a gambling casino in the middle of town, wouldn't be much different, except that tax money could be spent on local development and a casino would keep local people employed. Furthermore, advocates for a parasitic religion will work hard to expand the poor social conditions keeping it alive. They will cry out warnings about how insecure life and liberty is, they will point the finger at "dangerous" foreigners or "deviants" among us who must be combated, and they will try to weaken government's support for education, health care, and similar vital needs.

Insecure social/economic conditions and weak government together produce the social fragility permitting parasitic religion to supply the kinds of unnatural consolations that many anxious people want. Recent trends in America illustrate this as well as anywhere else. Economic conservatism would not ordinarily wish to ally with parasitic conservatism, since a large, secure, and educated workforce promotes the interests of industry and business. However, the wealthy class can find alternative ways to get even richer. Notable ways include war profiteering in collusion with governments starting unnecessary conflicts, sending jobs overseas where labor is kept cheaper by repressive governments, or financial speculation thanks to governments expanding the monetary supply. Eventually, this wealthy class doesn't have to care if dramatic gaps widen between themselves and everyone else. Parasitic religion can be a very useful partner in maintaining exactly the sort of social order beneficial to the extremely wealthy. A parasitic religion teaching that god approves of warlike aggression, expects individuals to be self-reliant, and smiles on prosperity however acquired, will especially find approval among both the contented aristocrats and the desperate poor.

Religious liberty cannot explain the origins of popular demand for religion, but a closer examination into adverse social conditions can. Religious leaders are always eager to support social order by allying with religious governments to together form an aristocratic class ruling over the people. Where a religion cannot control a secular government, it tends to resist rising equality and stronger social justice with staunch conservatism. The weaker a religion looks when compared to a secular government, the louder the religious cries are heard calling out for a comforting return to traditional ways.

Notes

1. Confucius. *Analects: With Selections from Traditional Commentaries*, trans. Edward Slingerland (Indianapolis, Ind.: Hackett, 2003), p. 20.

Further Reading

Bandyopadhyaya, Jayantanuja. *Class and Religion in Ancient India.* New Delhi, India, and New York: Anthem Press, 2007.

Bellah, Robert. *Religion in Human Evolution: From the Paleolithic to the Axial Age.* Cambridge, Mass.: Harvard University Press, 2011.

Berryman, Phillip. *Liberation Theology: The Essential Facts about the Revolutionary Movement in Latin America and Beyond.* New York: Random House, 2013.

Besio, Kimberly Ann. *Three Kingdoms and Chinese Culture*. Albany: State University of New York Press, 2007.

Chakrabarty, Bidyut. *Social and Political Thought of Mahatma Gandhi*. London and New York: Routledge, 2006.

Cline, Eric H., and Mark W. Graham. *Ancient Empires: From Mesopotamia to the Rise of Islam*. Cambridge, UK: Cambridge University Press, 2011.

Finer, Samuel E. *The History of Government from the Earliest Times*. Vol 1, *Ancient Monarchies and Empires*. Oxford: Oxford University Press, 1998.

Frankfort, Henri. *Kingship and the Gods*. Chicago: University of Chicago Press, 1948.

Garrow, David J. *Bearing the Cross: Martin Luther King, Jr., and the Southern Christian Leadership Conference*. New York: Perennial Classics, 2004.

Harris, Ian. *Buddhism, Power and Political Order*. London and New York: Routledge, 2007.

Hsü, Leonard Shihlien. *The Political Philosophy of Confucianism: An Interpretation of the Social and Political Ideas of Confucius, His Forerunners, and His Early Disciples*. London and New York: Routledge, 2013.

Irani, K. D., and Morris Silver, ed. *Social Justice in the Ancient World*. Westport, Conn.: Greenwood, 1995.

Kirsch, Jonathan. *God Against the Gods: The History of the War Between Monotheism and Polytheism*. New York: Penguin, 2005.

Michaels, Axel. *Hinduism Past and Present*. Princeton, N.J.: Princeton University Press, 2004.

Molloy, Michael, ed. *Experiencing the World's Religions*, 6th edn. New York: McGraw-Hill Higher Education, 2012.

Norenzayan, Ara. *Big Gods: How Religion Transformed Cooperation and Conflict*. Princeton, N.J.: Princeton University Press, 2013.

Paul, Gregory. "The Chronic Dependence of Popular Religiosity upon Dysfunctional Psychosociological Conditions." *Evolutionary Psychology* 7 (2009): 398–441.

Puett, Michael J. *To Become a God: Cosmology, Sacrifice, and Self-divinization in Early China*. Cambridge, Mass.: Harvard University Press 2002.

Quigley, Declan. *The Interpretation of Caste*. Oxford: Clarendon Press, 1993.

Trigger, Bruce G. *Understanding Early Civilizations: A Comparative Study*. Cambridge, UK: Cambridge University Press, 2003.

Vesselin Popovski, Gregory M. Reichberg, and Nicholas Turner, ed. *World Religions and Norms of War. Tokyo*: United Nations University Press, 2009.

Wiesehofer, Josef. *Ancient Persia: From 550 BC to 650 AD*. London: I. B. Tauris, 2006.

9

SCIENCE

Science for too long was fearful of its older sibling, religion. Both science and religion are born of wonder, like philosophy, but science has long experience with bowing and cringing and accepting compromise, anytime religion erupts into outrage over a new discovery about the world or humanity. This hostility is regrettable, since science and religion are both cultural artifacts, powered by the root drives of widening curiosity and sufficient explanation. Neither science nor religion lies closer to some "essence" of human nature. Science and religion grew apart because they adopted different methodologies for dealing with real human problems.

Religion tackled the complexities and problems of the social world long before writing and well ahead of science. Religion essentially remains tied to thinking about social relationships—between gods and nature, gods and humans, humans in society, and tensions within each person. The supreme virtue for the religious mind is harmony, and the supreme problem is disharmony. It follows that compelling religious traditions tend over time to design elegant theological narratives pointing to comprehensive solutions to all kinds of disharmonies. Philosophy emerged next, in the chaotic aftermath of early Iron Age clashes of civilizations and religions, and it retained the ideal of harmony as a supreme intellectual aim and guide to truth. Several of the

276

earliest philosophical intellects, from Greece and Egypt to India and China, hit upon the notion that supreme harmony must not depend on anthropomorphic agents or powers with personalities. The idea slowly emerged in several civilizations that the world's hidden energies are regular and organized but not sentient or even alive. The idea of science was slowly born.

Science and Religion's Natural Differences

Western naturalism is primarily indebted to the boldness of Greek rationalism and protoscience. (The rich traditions of Indian materialism and Chinese nature philosophy cannot be discussed here for lack of space.) The origins of natural science came from theorizing about what things are ultimately composed of, and what really controls nature's changes. In this new scientific way of thinking, more complex things are to be explained in terms of simpler things, and unpredictable events are to be explained in terms of more predictable regularities.

Greek speculations about earth, fire, water, and air, and other such "elemental" and atomic entities, attempted to simplify the picture of the basic parts of the world, and to explain natural changes in terms of habitual ways these basic parts can mix and combine together. Greek philosophy's origins are exemplified by Thales' speculation that all began with water and everything is ultimately made of water. The influence of religion may be involved here. Ancient religions from Egypt to Sumeria had identified a primeval source of water as essential to the story of the world's creation and sustenance. Yet Thales picked out water for its essential properties, its capacity to exist behind and within visible phenomena, and its ability to explain worldly phenomena as a necessary and lawful consequence of those properties—indicating that logical thinking was at work, not mythological. The Greeks were not the only thinkers to initiate logical and scientific thinking, but they speedily went the furthest among all early civilizations.

Religious thinking also seeks the hidden behind the visible, but it applies the complex ways that humans understand each other when attempting to understand the world's ways. For example, religions about animal spirits able to help us or demonic forces eager to harm us are essentially about careful negotiations that must be undertaken in special ways to preserve oneself and one's community. Religions with divine creators, angelic visitations, and inspired prophets are essentially about crucial communications and miracles that must be understood by everyone. What religious mythologies have in common is their focus on complex and unpredictable events happening at special times to privileged peoples. Such anthropocentric (human-centered) thinking is highly unreasonable when applied to the environing world, since it privileges the human perspective all out of proper proportion to nature's own ways. Instead of privileging one perspective conveniently favorable to a human community, natural science tries to offer explanations that can work from anywhere and be understood by anyone. Simple things (at least simpler than agents) and predictable regularities, confirmably valid anywhere in the universe no matter who you are, are the realities which science seeks.

Greek philosophies took three distinctive approaches to religion, and these options still supply the three primary options today. First, there are philosophies that compromise with religion by retaining divine gods and unearthly spirits as essential parts of reality. Second, some philosophies compromise with religion by postulating fundamental powers within nature to explain everything, and regard them as intelligent gods or guiding spirits. Third, there are philosophies that uncompromisingly reject anything unnatural/spiritual in their accounts of the basic powers and ultimate components of reality. These three approaches were bequeathed to their intellectual heirs, and they now define modern naturalism's difference from religious thinking. Do anthropomorphic deities, having their own personal characters and agendas, operate independently from the natural order and influence it as they may choose? This is the question of personality, and a naturalistic philosophy denies that anything like a person is responsible for the creation or maintenance of reality. Do thoughtful powers, having designs and plans, operate within the natural order to produce intended results? This is the question of mind, and a naturalistic philosophy cannot confirm that anything like a mind is responsible for the lawful structure of reality. Finally, we can ask about purpose. Do vital energies, having influences over matter, function within things to help produce and guide life? This is the question of spirit, and a naturalistic philosophy denies that the basic elements and fundamental laws of reality are in any sense alive.

*Aw c'mon... Can't you just believe
a little bit? For me?*

Western naturalism competitively evolved in opposition to religious thinking. Instead of a cosmogony—a tale about how god(s) intentionally made the world—science offers a cosmology describing a historical series of natural events that happened to make the world. Instead of discerning intelligence in the laws of nature, naturalism understands natural laws as only blind and heedless forces having no aim or purpose. Instead of postulating living sensitive spirits within life, science is learning how life chemically emerged from complex organizations of atoms that have no life of their own. Naturalism doesn't take agency as seriously as religion, because scientific method cannot rely so heavily on agency as an explanatory cause. Science can recognize agents as causes—mice surely cause the disappearance of cheese in their mazes—but science carefully restricts the need to appeal to agent causes in the world.

Most religions acquire their "explanatory power" by appealing to unnatural beings who behave like persons, in order to explain disrupting events and try to entice these personlike agents to help us. Science generally does not prioritize postulating beings with personal or agent-like abilities in its theories. The biological, psychological, and social sciences must deal with animals as agents, of course, but scientific method itself prefers hypothesizing nonagents to explain what is going on within organisms. Science prefers postulating things that are more regular and predictable, because that's what the abductive logic of testing hypotheses requires: to rigorously test a hypothesis, specific predictions must be made and precisely confirmed every time, so postulated entities must behave the same way under the same preset conditions. Chemists the world over, for example, must be able to confirm experimental results for themselves without worrying that their molecular compounds might surprisingly shift their preferences for combining with each other. That is why science must use regular impersonal entities and forces in explanations. For example, each atom has its definite characteristic properties, and atoms can't deviate from their natures or decide to misbehave. Atoms wouldn't be classifiable into elements otherwise—an atom of an isotope of carbon, for example, must be indististinguishable from and interchangeable with any other atom of that same carbon isotope (made of the same number of protons, neutrons, and electrons), as far as the formulas of organic chemistry are concerned.

Aside from this self-limitation by scientific method to rely on impersonal and nonagent explanations in hypotheses, there is no preset limit to what sciences may attempt to postulate. The experimental confirmation stage of testing hypotheses treats them all most severely and most hypotheses are weeded out as uncomfirmable and nonexplanatory, but the imaginative stage of hypothesizing entities can daringly propose ideas never yet conceived before. Such flights of imagination are the engine behind science's explanatory power by trial and error, as humans confront a staggeringly vast natural world having innumerable aspects and energies.

Science undertakes a quite different methodological project than religion's for explaining nature. Theologians and religious historians can promote simplistic and misleading views about science-religion compatibility, to be sure. Among atheism's advocates, some rely on mistaken and confusing stereotypes about science and religion as well. The following list illustrates such views. Each one makes a fair point about religion, but it also suggests a misleading contrast between science and religion.

- Religion imagines that there is a deeper hidden reality, while science only deals with what is observable.

- Religion projects wishful thinking onto an unknown reality, while science passively accepts the facts from a known reality.

- Religion relies on flights of fanciful imagination, while science relies on calculating rationality.

- Religion is imaginative and unconstrained by common sense, while science conforms to expectations of common sense.

- Religion proliferates into innumerable stories about unnaturally spooky things, while science talks about fairly ordinary natural things.

- Religion appeals to any numbers of strange powers to be dealt with, while science only uses one strict form of lawful explanation.

- Religion thinks that we can relate to higher powers to help us, while science thinks that higher powers are utterly independent and beyond manipulation.

- Religion thinks it guides people towards a hopeful future, while science tells people what will necessarily happen anyway.

- Religion will never converge on one story for all the answers, while science will eventually have a comprehensive explanation for everything.

These stereotypes about science cannot stand unchallenged and uncorrected. Here is a different list, better able to distinguish religion from science.

- Science does propose that there is a deeper reality behind what can be observed. Science searches for the hidden underlying energies and forces responsible for the behavior of observable things.

- Science does not know what reality is like in advance. Scientists must actively project their best guesses about what is presently unknown to find out what may be correct.

- Science relies on flights of fanciful imagination even more than religion, since religion is typically quite limited about imaginative ideas. Science must occasionally try radically new ideas to make progress.

- There is no limitation upon scientific imagination. Sciences occasionally postulate wildly counterintuitive and "unnatural" things that violate both common sense and what has counted as natural according to older scientific knowledge.

- Science doesn't settle for ordinary nature. The sciences postulate very different kinds of things. Physics describes atoms while biology deals with cells, and anthropology's cultures are very different from geology's continents. The quantum level of nature is quite different from our human-scale level of nature.

- Science's methods of inquiry have some commonality (there is no single methodology), but the use of scientific "laws" isn't monolithic. Some sciences use mathematical laws expressed by partial differential equations, some sciences only use statistical-probability laws, while other sciences can at best postulate habits and tendencies.

- Science understands nature through manipulating our interactions with it. Hypotheses are confirmed by demonstrating reliable control over anticipated observations and experimental outcomes. Passively collecting pure starlight, for example, isn't enough—instruments break starlight apart, sort frequencies and energies, and judge their significance and implications.

- Science does inspire visions of possible futures, and even better, it figures out how to really achieve them. Science helps people to technologically produce what they want and need, as they manipulate nature into one future rather than another. Where there are limits to our powers, science also informs us what they are.

- Science will never have a complete explanation for everything. For any "final" laws or set of "initial" conditions, science can always ask, and try to investigate, why just those laws and those initial conditions. "It just happened that way" is a halting of intelligence, not a scientific answer.

Scientific theories often are radically counterintuitive and genuinely novel, but not so speculative that they require more hypothetical agents than necessary or go beyond potential empirical testing. Philosophy of science is the field where the toughest questions about scientific methodology and theoretical commitment are explored. The differences between science and religion concern essential matters: what evidence they select, how they propose and choose explanations, and which kinds of accounts they find satisfying. Science and religion are especially divergent when it comes to the types of explanations they prefer to use. Science postulates predictable impersonal things (like atoms or forces) to explain general regular patterns, while religion typically postulates unpredictable willful agents (like spirits or gods) to explain particular extraordinary events. Religions seek cosmic explanations about otherwise mysterious and uncontrollable (and deeply distressing) events. By contrast, science seeks explanatory power within empirical limits and does not offer placebo rituals where no control is really possible.

Science and naturalism are often accused of denying that agency and personality really exist, just because scientific method refuses to postulate more of those things than absolutely necessary. But science makes no such denials, and neither should naturalism. Explaining how complex things like agents can exist and do what they do is not the same thing as explaining that those things were never real in the first place. One might as well argue that the clock maker who can explain all of a clock's parts and how they work together is thereby put in the unfortunate position of having to deny that any clocks have ever really existed. Throughout its history, science has attempted to provide sufficient explanations for natural things without resorting to adding more things having their own lives, minds, and fickle personalities. Living organisms, sentient life forms, and animals with personalities do naturally exist—we ourselves are obvious examples. However, none of these things are essential to nature's fundamental ways, and all of them are entirely dependent on nature.

That is why the physical sciences—physics, astrophysics, cosmology, chemistry, geology, etc.—are more fundamental than the life sciences, and why the life sciences are more fundamental than the social sciences. Personality depends on life, while life depends on physical energies, but never the reverse. The life sciences automatically refuse to consider hypotheses about things having personal intentions or purposes at the level of chemical metabolism or genetic reproduction. Similarly, the physical sciences automatically refuse to consider hypotheses about living or spirited things inside matter, or behind forces, or beyond nature. These scientific refusals are not equivalent to philosophically declaring that nothing mental or alive really exists, or that everything existing at a more complex level can't "really" exist and must instead be illusory, as if we suffer from some massive cognitive illusion or hallucination when we observe animals and people.

Nor are these scientific refusals to prioritize agent postulation just about stubbornly assigning an illusory status to everything that religions talk about. These refusals are not based on some built-in rejection of religious ideas, as if every new scientific theory has been only designed to contradict some religion or another. Rather, scientific explanations seek to satisfy the fundamental demand of sufficient reason that good explanations reduce mystery and reveal order, instead of increasing mystery and arbitrariness. The abductive realism to scientific method satisfies the postulate of sufficient reason, fulfilling the basic mission of science. This mission automatically sends science down a different explanatory path than that of religion. As religions continue to invent and multiply the number of complex spirited beings in order to explain already complicated events, no reduction in mystery is ultimately possible, since the powers and motives of those additional beings now need to be explained, and so on. The fact that religions simply stop at some point when narratives end, self-satisfied in their dramatic cast of unnatural characters, doesn't signal satisfactory explanation. We like good stories, to be sure, but a great narrative doesn't set down a valid standard for reasonable explanation that can work for any and every rational intelligence.

Why Should Science Accommodate Religion?
For its part, liberal theology has gradually accommodated science, in its slow but persistent way. But science has never accommodated religion, except in a limited political sense, and every "compromise" has been forced on science. For example, let's sketch out a selective history of some compromises between science and Christianity.

- A sixteenth-century "compromise" let science track mathematical relationships in the natural world. A growing revival of Greek ideas compelled theology to admit that nature could be understood by the human intellect. Any scientist interested in the notion that natural things can have their own inertias, energies, or laws, independent from god's direct guidance, was regarded as a heretic or worse: an "atheist."

- A seventeenth-century "compromise" granted permission for science to discern god's design of special creation. Science was expected to directly support natural evidentialist theology, so science had to discover how "perfections" in nature signal unmistakable evidence of divine planning. Any scientist discovering nature's oddities and imperfections, such as Galileo viewing sunspots and moon craters with his telescope, was labeled a threat to the Church.

- An eighteenth-century "compromise" was the Enlightenment's deistic notion that science learns how god's creation has been lawfully working (perfectly) since its origin. Science was expected to support inferences establishing the existence of a divine Creator, showing how nature's harmonies are so "convenient" to humanity that divine benevolence must have been at work. Even startling events like comets or earthquakes must be divine signs. Any science starting to show more interest in naturally unguided or chaotic forces, such as geology, was declared a threat to theology.

- An early nineteenth-century "compromise" was the Enlightenment's dualistic notion that science only has authority to explain matter, while religion explains spirit. Science was expected to track the measureable forces and energies of nature, but science could know nothing of god, the soul, the afterlife, or morality. Any science daring to explore consciousness or explain life, such as psychology and Darwinian evolution, is viewed as a threat to religion and its cultural dominance.

- A late nineteenth-century "compromise" with Darwinian biology admitted that the Creator used evolution to progressively produce humanity, the pinnacle of creation. Science was expected to confirm this epic narrative of inevitable progress, so long as gun-wielding white males can occupy the final highest position. Any emerging science charting the world's diverse cultures, their equal humanity, and their worthy accomplishments, such as anthropology and history, was regarded as a relativistic threat to Christianity's manifest destiny.

- A twentieth-century "compromise" offered by liberal Christianity shrank the role of religion to just social ethics, asking science to study only what is, leaving what ought to be for religion. Science was expected to supply knowledge about nature, society, and human psychology to serve as means for reaching the ethical goals set by religion. Any science serving a different master, such as behavioral psychology serving secular organizations, was denigrated as an existential threat to Christianity's core humanist values.

- A twenty-first-century "compromise" by popular Christianity so far sounds like, "We're just as mystified by god as nonbelievers, but that's the fun of faith, and Jesus is your best friend." Science is expected to confirm the psychological and physiological benefits of faith, forgiveness, meditation, churchly socializing, and so on. Any science finding how religion's effects are obtainable by nonreligious alternatives, such as neuroscience's stimulation of religious experiences, is categorized as a hostile reductionism eliminating genuine meaning from life.

This brief but not inaccurate sketch lends a certain perspective on "compromise." Reviewing such a series of grand compromises, it becomes clearer why friends of religion keep expecting science to reconcile itself with religion. Reconciling two parties does look easy when only one party can dictate the terms of reconciliation.

Overblown talk of a "war" between religion and science during the course of European history must finally be replaced by a nuanced appreciation of historical complexities. Partisans of atheism, no less than partisans of science, use metaphors about war too freely, down to the present day. The idea of war implies an open conflict between well-matched adversaries. Nothing of the sort was possible here, however.

First of all, there couldn't have been open conflict between religion and science for almost all of Western history since the "Dark Ages." The Christian Church controlled access to education for over ten centuries, including which books survived, what books could be read, who could be a educator and scholar, and who could be legally punished for freethought and heresy. From the fifth to the seventeenth centuries, clerical leadership enjoyed enormous powers within society and government, far greater than any power available to philosophers and scientists.

Very well, you may have math. You just can't do anything with it.

The myth that Christianity "saved" civilization and human learning really can't survive confrontation with factual history. The Church copied and preserved only what it wanted to, and left the rest to rot or burn. The Byzantines of the East managed to preserve many items that the West couldn't, such as Plato's and Lucian's complete works. Islamic academic centers saved an additional small fraction, to their credit. What finally survived down to present day from two large civilizations, all the remaining Greek and Latin writings put into printed volumes and brought together, would barely be able to fill a dozen bookcases.

Thinking about what did survive must cause one to appreciate pure luck. It is astonishing to grasp how many important classical works barely escaped complete erasure from the pages of history. The course of European history could have been very different if the Church had been even more neglectful. For example, just three copies of Lucretius's masterpiece *De Rerum Natura*, On the Nature of Things, survived. That book dramatically invigorated the Renaissance with practically its only materialist science and philosophy, alongside Aristotelian philosophy (especially Averroës) that arrived from Muslim regions, which could stand as an alternative to religion-infused thought. To next think about what was lost forever is to contemplate a horror and a tragedy for humanity.

How Did Science Survive Despite Religion?

Religion's friends like to claim that science got started because of help from religious ideas and religious institutions. The true story is that science rarely got much genuine help, and struggled to barely survive despite Christianity.

Church-controlled monasteries, cathedrals, and universities do have to receive credit for being the only place where almost everything of value was stored from the past. After all, the Church had systematically prevented any secular alternatives for any learning. Indeed, scattered and ill-funded Christian institutions ended up as the sole repositories for anything from the Greek and Latin worlds destined to survive in Europe. The fate of the brilliant lights from those proud worlds was in the hands of primitive half-literate zealots for their faith unable to appreciate the ideas of thinkers considered to be god-forsaken pagans who died to join the damned. That same monopolization on learning and thinking also allowed the Church to take what little credit it deserves for being the only institution where Greek and Latin ideas about the regular and mathematical nature of the cosmos could survive to invigorate the emerging empirical sciences of astronomy, physics, and medicine during the late Middle Ages.

The heretic Galileo, we are reminded by Christian historians, advanced scientific ideas already hinted at by late medieval philosophers, so some due credit must be assigned to Christianity. But the wider cultural context mustn't be ignored, either. It is the case that Galileo's use of inertial momentum to explain the motion of physical bodies was anticipated by Jean Buridan at the University of Paris during the mid-1300s. Buridan used the concept of impetus to explain why an object, once it had acquired a motion in a direction, would continue to move that way unless something else impeded it. Buridan understood that he was departing from Aristotle, but fortunately for him, he had avoided joining a religious order and becoming a theologian, and he died before charges of heresy arose. His writings were prohibited a century later, when a spasm of theological orthodoxy suppressed nominalist and mechanistic speculation, so progress in physics was delayed for two more centuries until Galileo. But Buridan was not the first Christian philosopher to contemplate impetus.

Between Aristotle and Buridan, there stood John Philoponus, a sixth-century thinker, who proposed a version of impetus and opposed Aristotle on other grounds as well. Yet Philoponus couldn't inaugurate a school of experimental philosophy, and nothing of his scientific reasonings was taken seriously until seven centuries later. The reason is simple: he had converted to Christianity, got entangled with metaphysical debates over Neoplatonism and the Trinity, found his challenges to Aristotle unappreciated, died a lonely figure, received the posthumous condemnation of "heretic" a century later, and therefore he stood as an ominous warning to Christian scholars for many centuries thereafter. Let the life of Philoponus illustrate how Christianity treated independent philosophical thinking and the beginnings of scientific reasoning. When historians are so eager to neatly tie "Christianity" and "science" together in the same sentence, the full story hasn't been revealed. Yes, science and religion in Europe had to cohabitate the same intellectual spaces for many centuries. Yet no credit at all should be bestowed for mere happenstances of time and situation. Science isn't Christian. If any secular alternative had been able to coexist alongside the Church, preserving even 10% of classical learning and protecting scientific inquiry, the intellectual progress of Europe would have been strikingly different.

Instead, empirical science and experimental technology was effectively kept comatose for ten centuries. When they finally awoke, they were still hostage to an alien religious host and took what little sustenance was available. When a historian of science and religion can only offer the compass, paper, printing, gunpowder, and cannon in praise of the "enormous" advances across a thousand years of Christianity (and then promptly admits how they were all borrowed from Asia), one rightly suspects that despite what this historian proclaims, science couldn't have been a Christian priority. And when this same historian (he is James Hannam) can't mention any significant advances in theoretical science until after the resurrection of the Greek body of thought in Europe, a resurrection that the Church didn't initiate or encourage, then one cannot be faulted for wondering what credit for science Christianity really deserves.[1] Finally, if one needs to be reminded one more time that modern science followed medieval science, just as the seventeenth century had to come after the twelfth century, then forgetfulness about modern science's courageous departures from Christian dogma can be excused. One might as well argue that since democracy's emergence in the eighteenth century came after fourteenth century theology, then medieval Christianity deserves some credit for democracy. Or one could somehow argue that since all the founders of democratic thought happened to be Christians, then Christianity had to be democratic. Yet the bold advocates of constitutional democracy were not reading the Church Fathers—they were reading Greek and Roman political theory, such as Aristotle, Cicero, and Seneca, searching for ways to replace the rigid traditions of Christian thinking.

Although nothing properly Biblical or Christian could have inspired scientific ideas about the natural world having its own energies and laws, religious historians still try to depict the emerging protosciences of the thirteenth and fourteenth centuries as fundamentally reliant on Christianity. Early science's continued heavy reliance on Christian theology shows only how well Christianity nearly erased the direct Greek sources from view and left religious institutions as the only places to access any of those ideas. Not until the twelfth century did Latin translations of Arabic learning reach Europe, and Latin translations of Plato's entire works had to wait until the fifteenth century. Almost no one could read Greek in all of Europe until the dawning of the Renaissance, and classical Latin works remained extremely scarce.

If the early sciences were inspired by some theology, that theology about a natural world of order and lawfulness was borrowed from the Greeks. The greatest influences were Plato's book Timaeus, along with Stoicism, which was crucial for the early Church, and then Aristotle, who was essential to the medieval Church. To its credit, Christian theology was always self-consciously eclectic, having incorporated many aspects of pagan thought, including the concepts of natural law and civic polity, for its second- and third-century foundations. By the late medieval and early Renaissance periods, scientific and philosophical innovators became even more eclectic, drawing upon components of Plato, Aristotle, Cicero, and Aquinas, along with the Stoics, Aristotelian commentators, Averroës, Duns Scotus, and many more sources. Religious historians only pointing to Christian theology's ideas about such things as natural laws and "reading" the signs of nature aren't reliable about properly crediting Greek and Roman thinkers. Most of those thinkers and their philosophies were amply religious, especially the Platonists and Stoics, but they weren't Christian and their pagan gods had to answer to philosophy, not the other way around.

Much of pagan religion and philosophy was lost to Christian Europe, causing so much forgetfulness. It must be reemphasized how access to original Greek and Roman thinkers was extremely limited after the fall of Rome. Most of the Greek and Latin writings that even survived in Europe were hardly read or used for all those centuries, scattered across the countryside of Europe as small monasteries were. In the cities, to think about the natural world at all was to think about it through many filters of theological reinterpretation and religious dogma, with only limited and partial assistance from non-Christian sources. As eclectic as brilliant minds attempted to be, they all had to ultimately work within some Christian theological system or another. Early science had no choice but to think about the natural world with influential theological categories in mind, since there was no non-religious alternative available. On this matter, all historians must be agreed. The deeper questions are whether science would have emerged without Christianity, and whether Christianity deserves praise for science's dependency. The correct answers are these. On the first question, science would have emerged centuries earlier without Christianity's domination, especially if Greek and Roman science had been more accessible. On the second question, Christianity deserves very little credit at all. Early science would have been vastly better off continuing to grow under the gentler worldviews of materialistic atomism, Stoicism, or naturalistic Aristotelianism than the heavy yoke of providential and autocratic monotheism. To imagine that science was truly better off with Yahweh is to imagine that if somehow science had to start over again today, the finest place in the world for that rebirth would be Taliban-controlled Afghanistan.

Religious historians of "Christian" science never fail to tell well-crafted stories about the first important scientists. They can't fit the heretic Galileo into their pleasant narrative, even after correcting the "Galileo in prison" exaggeration, so they prefer to talk about more docile Christians such as Isaac Newton and Johannes Kepler. Both of those scientists were motivated to understand the divinely mathematical laws of motion, just as so many thinkers had attempted since the termination of classical science over a millennia before them. But neither of these thinkers was a creature of the theology schools or Church-run universities, and they held distinctly unorthodox religious views. In England, Newton discovered gravity and the universal laws of motion by boldly departing from dogmatic Christian metaphysics along with the Christian worldview requiring separate laws of motion for the Earth and the heavens. His basic discoveries took place away from Cambridge University as a young man; in later years, Newton could hold his mathematics professorship, despite the fact that he wasn't an ordained priest, only by special decree from King Charles II. In Germany, Kepler avoided theology, apprenticed himself to independent astronomer Tycho Brahe, and became the court mathematician of Emperor Rudolf II. His astronomical data was better than anyone's, to be sure, yet he determined how the planets track elliptical orbits, succeeding where all previous Christian astronomers failed, only after he set aside Christian dogmas about the sun going around the Earth and the planets following circular orbits. Scientific geniuses during those centuries may have been motivated by thoughts of god, but they contributed to science by abandoning obstructions from Christianity.

If theology and science were viewed as generally "compatible" (scientific genius excepted) for century after century, it was because theology controlled thinkable worldviews for almost every intellectual. Let no one imagine that the finest centers of Christian learning were rich with classical literature. Even the University of Paris of the fourteenth century, the greatest center of Christian learning in Europe, had a library of perhaps just 2,000 volumes, the bulk of which were about the Bible and religious subjects. A genius did erupt there at Paris, the great Aquinas who was capable of digesting Aristotle, but his theological synthesis subverted Aristotle's natural philosophy into metaphysical theology in service to the Bible's God. Long after the reemergence of nontheological and scientific alternatives during the Renaissance, along with fresh translations of Islamic and Jewish works and additional Greek thought arriving from Byzantium by the 1400s, their radical and "heretical" nature kept them marginalized. These marginalized works nevertheless had powerful effects during and after the Renaissance, as free thought began to challenge orthodoxy. Two writings from the days of the Roman Empire were especially valuable. Cicero's *On the Nature of the Gods* taught Europe how to debate whether gods exist, and Sextus Empiricus on Pyrrhonism taught bold thinkers how to be skeptical about almost anything. Similarly, during the Renaissance, the Church's schools and universities were the only place where the new Humanism could be studied and enlarged, but very little of that new way of thinking spread beyond city centers, and education in general was hardly affected.

European universities remained for the most part conservative bastions of orthodoxy. The occasional dissent against scholasticism or trinitarianism were exceptions proving the rule. Little original scientific research took place there. Medieval Aristotelianism, intelligently designed by theologians to require a providential god, remained the prescribed worldview for curricula across Europe and America until the nineteenth century. Defenders of Christianity can at least point to the existence of universities, so central to the development of European civilization. The first universities, and all universities until the late nineteenth century, were indeed Christian. Of course, there was no other alternative, and the Church does not deserve as much credit as might be supposed. Most of the great universities got started and evolved with only modest Church involvement. The University of Bologna was founded by legal scholars working on civil and secular issues, studying Roman law as much as Church canon, and it flourished under Emperor Barbarossa's protection by the mid-1100s. The University of Paris must not be confused with the ecclesiastical school at Notre Dame Cathedral. The masters of the new university attracted their own students for studies in liberal arts, medicine, law, and theology, and only later were they gradually brought under Church supervision. Oxford University grew up where administrative councils and juridical courts attracted scholars, and as more masters began teaching liberal arts and law, they were later joined by theologians.

Most of the universities started during the twelfth and thirteenth centuries were founded by visionary kings and civil authorities in need of lawyers and bureaucrats, and only later did the Church authorize their theology faculties. By the fourteenth century, the Church exerted doctrinal authority over all university faculties.

A Declaration of Science's Independence

Conformity was not hard to impose by the Church, but conformity needn't impose total stagnation. It must be reasserted some modest intellectual progress was happening by the fifteenth and sixteenth centuries. But where? Much original scientific research had to occur beyond the reach of church and university, undertaken by amateurs or scholars patronized by wealthy benefactors. Those without a well-connected and powerful patron were far more vulnerable to censorship and civil punishment. Where some original science was subsidized by the Church, the recipients were carefully scrutinized and selected for religious orthodoxy.

Governments assisted the theological goals of thought control and censorship through terror. The medieval Church could interrogate heresy with torture, but murder for blasphemy had to be done by the state. The maniacal pursuit and destruction of religious heretics deviating from orthodoxy was a grave concern to those privately harboring deistic, pantheistic, or materialist views. As the Renaissance began, the Inquisition didn't end for Europe's intelligentsia, and matters only grew worse. The period of 1450 to 1650, as empirical science and naturalistic philosophy were just getting started, witnessed severe Catholic oppression. There were no "secular" philosophers until Thomas Hobbes and Baruch Spinoza during the mid- to late seventeenth century, and there were very few genuine atheists to burn, but religious philosophers and scientists displaying any notable originality found their books banned and burned, and many were investigated by the Church. The most defiantly naturalistic among them became victims of the Inquisition. Pico della Mirandola had books banned, spent time in prison, and was treacherously poisoned. Lucilio Vanini was burned at the stake. Gerolamo Cardano was accused of heresy and removed from his professorship. Galileo was silenced for many years and died under house arrest. Tommaso Campanella spent twenty-seven years in prison. Giordano Bruno was imprisoned for seven years, tortured, and finally burned at the stake. It sounds like a relatively small number of victims, but their examples were well-publicized and quite sufficient for widespread intimidation.

By 1650 the worst was over. Prominent freethinkers after that date didn't die at the hands of religious or civil authorities. At that time, Christian scientist Pierre Gassendi espoused empiricism and atomism while holding a professorship at Paris, and he died of old age. Thomas Hobbes's *Leviathan* appeared in 1651, and he later died a natural death. Baruch Spinoza was excommunicated by his Jewish community for his *Tractatus Theologico-Politicus* (1670), and his *Ethica Ordine Geometrico Demonstrata* (1677) appeared after his premature death from lung disease. Protestants weren't any more receptive to heretical naturalism than Catholics, but Protestant countries tolerated scientific inquiry fairly well. Nevertheless, if every European scientist before 1700 was a public Christian (or was Jewish), that says nothing about the value of Christianity for science, but everything about the pervasiveness of theological thinking and the priority of keeping one's head attached to one's neck.

No, there couldn't be a war between religion and science, since religion had all the power and exercised it effectively. For many centuries, science had no option but to ally with theology just to survive better than the Greeks and Romans did.

Not until the eighteenth century, with rise of nation-states ruled by monarchs tolerating religious diversity and dissent and appreciating science's value for society and state, could many thinkers openly describe a physical world of natural laws as experimental science really finds it, requiring no divine involvement. As scientists enjoyed greater liberty of thought and publication, they sought a balance between science and religion, and the Enlightenment basically demanded that science play an equal partnership role. Deism and even pantheism arose as metaphysical options, and a handful of materialists were able to publish their views. The nineteenth and twentieth centuries then witnessed a surprising reversal of roles, with science growing dominant. This reversal wasn't welcomed by Christianity, to say the least. Great pressure was now applied upon religion to yield to science's powerful new knowledge, and theology had to hastily adjust its own cosmology at an unaccustomed pace. Still, it was fresh work for theologians, and many built impressive careers, including those heading back into fundamentalism. By the late nineteenth century, friends of science could keep their careers while complaining about science's long subservience to religion. The apron strings tying the maturing sciences to the Church's matronly figure were finally cut.

All the same, science's exultations over liberation from its captivity didn't have to be so impetuous, announcing victory in a war that it really never was able to fairly fight. This struggle's first chroniclers happened to be two Americans enjoying new academic freedoms in the New World: John William Draper published his *History of the Conflict between Religion and Science* in 1874, followed by Andrew Dickson White's *A History of the Warfare of Science with Theology in Christendom* in 1896. News of a war, much less a loss in that war, was widely greeted with disbelief by believers. Surely there'd been no war, as far as Christians could recall—only admirable conciliations and harmonies between science and religion! Negative reactions to Draper's *History* varied, depending on the denomination of the reviewer, but all reviews agreed that Draper had composed only fiction. Religious critics also agreed on one further point: if there's any war to speak of, it has arrived in the form of belligerent science. All the same, nineteenth-century science was only asking for independence, hoping that war wouldn't be necessary. Religion's defenders found themselves in a conflict of their own making, since they expected to continue dictating how the science-religion relationship would work.

Well-meaning peacemakers among today's friends of religion are sustaining those medievalist longings for science-religion harmony, asking for gentle "accommodation." However, science has no obligation to accommodate itself to religion, even if it could. Pointing to all those ways that science can "reconcile" or "compromise" with religion, across centuries and millennia, says nothing about whether science and religion are truly compatible. They are incompatible in methodology, incompatible in theoretical results, and they don't heed each other's goals. That's why metaphysical philosophies and revisionary theologies require so much creative effort to try to reconcile them. Those who say that science-religion compatibility has been easy are people already committed to an established metaphysics or theology dedicated to making it look easy.

Science and religion are belief systems relying on distinct methods—science requires experimenting with what is empirically accessible, while faith requires living in accord with what is absolutely

supreme. Religion won't seriously put faith on trial by empirical results, and science won't willingly place facts beyond empirical testing.
Ultimately, religion takes responsibility for envisioning how a congenial cosmic home can be maintained for humanity's hopes, while
science takes responsibility for theorizing how the environing world
maintains itself regardless of our hopes. Religion must have an axiology
about values built into the cosmos; science can deal with values (as later
chapters explain), but discerns no cosmic values.

Science vs. Religion Today

When atheists say things like "Science refutes god" or "More science
means less religion," it is not clear what is being claimed. Scientific
knowledge itself cannot disprove all gods, since theologies have relocat-
ed many gods safely beyond empirical reach. Atheism must take care.
Saying "scientific knowledge refutes religion" only arouses the theological reply that "science can't explore the supernatural, just the natural"
so that science looks quite incompetent to judge god's existence.

This natural science vs. supernatural theology "division of labor"
sustains the outdated Enlightenment divide between the little nature
that science explores and the big heaven that religion surveys. That
divide was laid down by theologians and religious philosophers during the seventeenth and eighteenth centuries as a defensive posture
against science. Catholicism and much of Protestantism gradually accepted science's free reign over nature by the mid-twentieth century,
although evolution delayed reconciliation for a long time. Natural
science's agreement that values and ethics can't be naturally compre-
hended was gratefully accepted by theology, despite the protests of social and behavioral sciences, which were promptly demoted to lesser
scientific status in a spasm of positivism and reductionism during the
mid-twentieth century.

To this day, the strange spectacle of scientists confidently knowing
where the reach of empirical inquiry must end makes a pleasing display of peace with theologians confidently knowing where the supernatural begins. Zoologist Stephen J. Gould, impressed by the Catholic
Church's gracious retractions of its condemnations of Galileo and

Darwin, went happily to his confession about how limited scientific knowledge must always be. His endorsement of the doctrine of "Non-Overlapping Magisteria" (NOMA) agrees that science studies natural facts while religion handles supernatural mysteries, and NOMA encourages science-religion cooperation. NOMA simplifies everything, because no longer must one ponder whether god-belief is even rational, or whether religion really ensures meaning and morality. However, atheism must be careful about the strategies needed for combating supernaturalism. Richard Dawkins' wise strategy says that something impossible to empirically test should be set aside as unbelievable. The problem with god is not that science hasn't delivered a verdict about evidence yet, but rather that there's never any evidence. No evidence, then nothing to think about—and surely there's no scientific question about it, and no scientific hypothesis to test. Since no one should believe that religion has knowledge of anything supernatural, religion has no magisteria, and NOMA collapses. Dawkins' criticism of NOMA is quite sound.

Nevertheless, friends of religion like NOMA, and they like to dictate the boundaries of science to scientists. We hear talk like, "Science only deals with natural events and observable matters." Such statements can sound commonsensical or even tautological, but they conceal hidden agendas unfavorable to science. When this statement mentions

"natural events," what would count as an unnatural event, and who would be able to detect them? And science surely deals with hypotheses about nonobservable matters with its methodology. Conveniently defining god as entirely supernatural (and unobservable?) lets religious intellectuals think they have erected a Wall of Separation between science and religion.

That Wall of Separation makes no sense, for religion or for science. Religion must constantly breach that wall so that god(s) can do something with the world, and science methodically ignores it so that new theories can expand what "nature" means. It is irresponsible to assume

that science and religion can be kept forever apart by some artificial wall. Religion itself is mostly responsible for tearing down the Wall of Separation as fast as theologians try to build it up (as previous sections track the history of religion with science). After all, religions invoke unnatural powers to affect people's lives, along with unnatural powers within people to enable immortality. Abstract deities unable to affect the human world couldn't remain interesting to practical-minded people. Real gods supposedly able to affect people's lives in real ways can hold human attention, as religions have always known. Science won't agree with religious expectations about divine interventions. From science's perspective, nothing can reasonably be attributable to unnatural powers.

God Is Not a Scientific Hypothesis

No god hypothesis should ever receive the sort of consideration that a scientific hypothesis should receive. Religious ideas about some supernatural creator responsible for the world only appears to enjoy vast empirical support. But they really can't enjoy any at all.

Crediting a god with creating something in the world or causing a miracle to happen in the world is far more like pseudoscientific "explanations." At least pseudoscience tries to look and sound like something scientific is proposed. However, pseudoscience is exposed by the manner in which a suggested explanation is suspiciously formulated and poorly tested (or not tested al all). Here are eight ways to determine that someone is proposing something that must fall short of scientific status:

1. An explanation is offered for phenomena that haven't been scientifically observed yet.

2. An entity is postulated that has no clearly defined identifying qualities.

3. An entity is postulated that would not be responsible for any unexpected natural patterns.

4. A postulated entity is so contrary to established scientific knowledge that experimental testing is impossible.

5. No predictions are expected to follow from the hypothetical existence of the postulated entity.

6. If some predictions are expected, they are either vague, difficult to experimentally test, or unsurprising.

7. Any predictions from the hypothesis that turn out to be false are simply ignored.

8. In case of an embarrassingly bad prediction, the hypothesis is modified afterwards to actually "predict" that different result.

Ordinary rationality agrees here with science. A speculative idea, like imagining that there is a god that will never permit anyone to see any distinctive sign of its existence, is paradigmatically unreasonable according to plain common sense.

The swift verdict of science is that no notion of a supernatural god could ever deserve any intellectual respect. The swift verdict of basic rationality is that deities (or any other mythical beings) deserve no more credibility than fictional characters in children's story. Not only must science reject gods, but any intelligent mind must also refuse to even consider them seriously. Science does not directly disprove religion, but it exposes why rationality itself must avoid it.

Notes

1. James Hannam, *The Genesis of Science: How the Christian Middle Ages Launched the Scientific Revolution* (Washington, D.C.: Regnery Publishing, 2011). A corrective to the biased agenda that Hannam represents was already published by Charles Freeman, *The Closing of the Western Mind: The Rise of Faith and the Fall of Reason* (London: William Heinemann, 2002; New York: Alfred A. Knopf, 2003).

Further Reading

Begley, Ronald, and Joseph Koterski, ed. *Medieval Education*. New York: Fordham University Press, 2005.

Bolgar, R. R. *The Classical Heritage and Its Beneficiaries*. Cambridge, UK: Cambridge University Press, 1973.

Brooke, John Hedley, and Ronald L. Numbers. *Science and Religion Around the World*. Oxford: Oxford University Press, 2011.

Carrier, Richard. "Christianity Was Not Responsible for Modern Science," in *The Christian Delusion*, ed. John W. Loftus (Amherst, N.Y.: Prometheus Books, 2011), pp. 396–419.

Dawkins, Richard. *The God Delusion*. London: Bantam, 2006.

Draper, John William. *History of the Conflict between Religion and Science*. New York: D. Appleton and Co., 1874.

Frazier, Kendrick, ed. *Science Under Siege: Defending Science, Exposing Pseudoscience*. Amherst, N.Y.: Prometheus Books, 2009.

Freely, John. *Aladdin's Lamp: How Greek Science Came to Europe Through the Islamic World*. New York: Random House, 2010.

Freeman, Charles. *The Closing of the Western Mind: The Rise of Faith and the Fall of Reason*. New York: Alfred A. Knopf, 2003.

Gaukroger, Stephen. *The Emergence of a Scientific Culture: Science and the Shaping of Modernity, 1210–1685*. Oxford: Oxford University Press, 2006.

Gould, Stephen J. *Rocks of Ages: Science and Religion in the Fullness of Life*. New York: Random House, 2002.

Greenblatt, Stephen. *The Swerve: How the World Became Modern*. New York: W. W. Norton, 2011.

Highet, Gilbert. *The Classical Tradition: Greek and Roman Influences on Western Literature*. Oxford: Oxford University Press, 1949.

Klima, Gyula. *John Buridan*. Oxford: Oxford University Press, 2008.

Kraye, Jill, Tom Sorell, and Graham Rogers, ed. *Scientia in Early Modern Philosophy: Seventeenth-Century Thinkers on Demonstrative Knowledge from First Principles*. Berlin and New York: Springer, 2010.

Lang, Helen S. *Aristotle's Physics and Its Medieval Varieties*. Albany: State University of New York Press, 1992.

Moore, R. I. *The War on Heresy*. Cambridge, Mass.: Harvard University Press, 2012.

Morvillo, Nancy. *Science and Religion: Understanding the Issues*. Malden, Mass.: Wiley-Blackwell, 2010.

Murphy, Cullen. *God's Jury: The Inquisition and the Making of the Modern World*. New York: Houghton Mifflin Harcourt, 2012.

Olson, Richard G. *Science and Religion, 1450–1900: From Copernicus to Darwin*. Baltimore, Md.: Johns Hopkins University Press, 2006.

Orme, Nicholas. *Medieval Schools: From Roman Britain to Renaissance England*. New Haven, Conn.: Yale University Press, 2006.

Peters, Edward. *Inquisition*. New York: Free Press, 1988.

Saliba, George. *Islamic Science and the Making of the European Renaissance*. Cambridge, Mass.: MIT Press, 2007.

Shermer, Michael. *Why People Believe Weird Things: Pseudoscience, Superstition, and Other Confusions of Our Time*. New York: Macmillan, 2002.

Sorabji, Richard, ed. *Philoponus and the Rejection of Aristotelian Science*, 2nd edn. London: University of London, 2010.

Styers, Randall. *Making Magic: Religion, Magic, and Science in the Modern World*. Oxford: Oxford University Press, 2003.

Weinberg, Steven. *Facing Up: Science and Its Cultural Adversaries*. Cambridge, Mass.: Harvard University Press, 2012.

White, Andrew Dickson. *A History of the Warfare of Science with Theology in Christendom*, 2 vols. New York: D. Appleton and Co., 1896.

10

TECHNOLOGY

Religions are obsessed with controlling the powers of people and the relationships between people. The enemy is change. Preventing personal empowerment and social change that might happen through access to new technologies has long been a high priority of religion after religion, all around the world.

We are all too familiar with the struggles between technology and religiosity. Both resisters and promoters pursue their righteous causes with intense fervor. Apocalyptic end-times are repeatedly forecasted, and then pushed on ahead further into the future after predicted deadlines pass. From the original nineteenth century Luddites who were protesting industrial production to the utopian transhumanists happily preparing for the twenty-second century, visionary idealism tries to keep pace with the never-ending flood of new technologies.

Humanity's End
Should advancing technologies be permitted to profoundly change our lives, and radically transform cultures? From communications (books, cell phones, Internet) and medicine (fighting disease and extending life) to the teaching of modern science, scientific technology and technical globalization have overtaken most of the world. Religions are confronted daily with the challenges of novel technologies. Different

religions have been able to adapt, to widely varying degrees, and often in surprising and fruitful ways. Religion and the Internet seem to be made for each other, letting new religions spread as fast as the speed of light, but old-fashioned creeds are exposed as outdated just as quickly. Every new technology offers great promises, and great perils. How do we deal with religious conservatism's worries over reproductive technologies, surgical "miracles," and the possibility of ultimate victory over death itself? Those worries seem foolish to many progressives. All the same, religious conservatives are not alone for worrying about radical technologies elevating the risk of potential violations against human dignity or equal value. Everyone should give some thought to disruptive and even destructive consequences of technologies.

Although religions are too quick to ignorantly slap a label of "evil" on some new technologies, that doesn't mean that technologies are always smart or reliably beneficial. Anything falling within the clutch of humanity's grip can be applied for doing good or evil, as good or evil as the mind controlling the application. Responsible minds must stay in control of technology, lest we permit technologies to gain control over us. Religions take a shortcut around responsible wisdom: instead of promoting thoughtful responsibility, they try to simply dictate what is "truly" good. Religions all suppose that they alone really know what is good for humanity in the long run, and they claim to know humanity's purpose and how it will all end.

If there is anything from the mind of humanity that gives us practical purposes in this life, it must be technology. From the Greek word "techne" to indicate anything crafted or constructed, our technologies are what we do best. Humans have designed and redesigned almost everything we can do with our primate inheritance. By the time that Homo sapiens had evolved from descendants of Homo erectus, our species was already reshaping the simple tools we could use and altering the environments we could reach. If the end of living is to have lived prosperously and well, then technology seems to be humanity's end.

The earliest human technologies long predate religion, by over a hundred thousand years. After religious practices were invented, religions applied all sorts of crafts for their own purposes. The practices

of human sociality, such as instruction, morality, storytelling, visual arts, singing, and dancing, are all social crafts far older than religion, and religion readapted them to better convey its messages. Significant advancements in technology as civilization developed, from writing and architecture to medicine and mass media, have been taken up by religions and put to religious purposes besides their mundane uses. So long as religions could figure out how to control new technologies for themselves and limit how most people could access them, religions can be kept satisfied. It is constant work, however. The big business of religion often feels threatened by new technology. It is difficult to name any other kind of social activity among human beings, besides religion, which always suffers when novel technologies are invented. And you can't name any social institution more likely to condemn a new technology than religion.

The technologies declared by religion to be dangerous and evil are too many to count. Think of just about any technology, and some religion somewhere has scorned it. The technologies earning the most religious contempt and fear fall into predictable categories. Religions don't like challenges to their control over people's public identities and their private minds, so religions resist new ways for people to commune together and communicate with each other. Religions particularly combat any educational methods and institutions that they can't control. Religions hate competition to their myths, so they fight fresh explorations and explanations for the world around us. Religions want to control people's own bodies, so they resist better ways to promote health and extend life. Religions like to regulate family structures and sizes, so they resist people gaining control over reproduction and pregnancy. Religions also love to intertwine themselves into the economic system of a country, wherever they can, in order to parasitically drain away resources and grow their own wealth. Religions especially tend to resist any economic developments threatening the wealth or privilege of the upper classes. Religions especially don't like novel techniques permitting people to regulate and govern their communities without the religious supervision of a clerical bureaucracy.

When it comes to practical technologies for farming and construction, the crafts of manufacture and commerce, or the weapons for war, a religion typically regards the technologies available at the time of its own origins as entirely natural and appropriate to use. When religions degenerate into fundamentalisms yearning for a return to the better times when God seemed closer, or when deities walked among us, then religions enshrine that earlier way of life as the only way to live. A few fundamentalisms, like the Amish, try to continue living much like their sixteenth-century ancestors did. Sects from some religions can still be found today trying to live like ancient Bronze Age peoples. Only religion has inspired people to live the impoverished lives of their distant ancestors.

Religion hasn't allowed any aspect of daily life to proceed without inspection and judgment. Adornments and displays, messaging and teaching, experimentation and learning, medicine and surgery, sexuality and contraception—the most intimate and personal aspects to our lives have fallen under the scrutiny of religions. There are no intrusions, private or personal, that religions aren't ashamed to venture into people's lives. This kind of broad impact is even greater than the power of police and politics. Religions aren't above infiltrating governments either, in order to use public officials and armed forces to enforce religious laws. Church and state wouldn't be so hard to separate if religions gave up their fanatical quest to use civil servants as their own religious servants. Yet that fanatical quest continues, perpetuating an entanglement of religion and politics going back to the dawn of civilization.

During the long period of slow technological advancement from 5000 BCE to around 500 BCE, most major religions were able to keep pace with technology. Technology was fostered and put to religious use. The Egyptians built the sacred mausoleums that we call the Pyramids, Babylonians raised up their ziggurats to the heavens, the Hindus mastered every artistic form including poetry and sculpture, and Chinese Daoism fostered an immense variety of health regimens and medicines. And every major religion was delighted to benefit from advanced weaponry to win their religious wars.

Dangerous Technology
As humanity gradually left the Bronze Age and entered the Iron Age by 1000 BCE, depending on the geographical region, technological change quickened its pace. Beginning technologies and the first early sciences had to struggle to liberate themselves from misuse and monopolization by religions. "Dangerous" technologies that have frightened religions over the past centuries include the alphabet, the book, the telescope, the ballot box, surgery, movies, vaccinations, and condoms.

The slow spread of reading and writing itself is a long, agonizing tale of gradual emancipation from religious institutions that obstructed the spread of literacy to the masses. The science of astronomy took millennia to escape servitude to the fraudulent practice of astrology. Cartography and geology were retarded for centuries due to religious dogmas about the shape, dimensions, and age of the Earth.

Thou treadest upon dangerous ground, brother Jacob...

Mineralogy and chemistry had to similarly find freedom from magical alchemy. Medical skills were so bonded to religious cults that any genuine medicinal value to an arcane remedy was obscured by a dazzling façade of priestcraft magic. Mental illnesses were "treated" by incantations and exorcisms; incurable physical troubles were explained away by pointing at the sufferer's sins. As truly beneficial technologies became available, one by one, with each slow step of intellectual progress, religions recoiled and protested because their lucrative scams were getting exposed and replaced. Where fundamentalist and fanatical religions control governments today, access to the Internet without restrictions or censorship is difficult to obtain. This amazing means of global communication, able to connect and educate billions of people, has become the one thing feared most by religious dictators, although secular dictators aren't far behind in their paranoia.

One of the oldest recorded religious prohibitions against the use of a technology is the ancient Jewish prohibition (pre–sixth century BCE) against the practice of tattooing, in Leviticus 19:28. Ancient religions from the Mayans of Central America to the Buddhists of Southeast Asia upheld religious rules about body ornamentation and appropriate adornment. Many religions today still control the public appearance and clothing of followers as well as the costumes worn by religious leaders. Banning the use of certain technologies by the faithful is a nearly universal feature of religions. Religious obedience simultaneously serves as a distinguishing marker to mark the truly righteous, and it keeps them dependent on religion. The Jewish practice of doing "no work" on the Sabbath can amount to a prohibition against utilizing almost all technologies on that day. The more traditional Amish communities avoid most technologies invented after the 1700s for religious reasons, especially anything using motors or electricity. Both Buddhism and Jainism have inspired religious communities that emphasize simplified and nonwasteful lives, amounting to an avoidance of many modern technologies.

Many more examples of long-standing prohibitions against technology can be found, too many to try to list. Here are two additional examples that deeply affected two major civilizations. The prohibition in the Hippocratic Oath against the use of poisons to induce euthanasia or abortion was probably related to religious ideas shared by the Hippocratic school of medicine dating back to the fifth century BCE. This odd prohibition, strange even in ancient Greece where both euthanasia and early abortion were tolerated, would have died out with the extinction of Hippocratic cults—except for the intervention of Christian theologians, who rediscovered these prohibitions and promoted them as Christian rules. A second example comes from India. Among the major Vedic works of ancient Hinduism, the Baudhayana Dharma Sutra strongly discourages ocean travel for the upper castes. This curious prohibition only slowed seafaring trade somewhat, since mercantile trade was mostly undertaken by lower castes anyway. However, it did limit exploration and cultural exchange for many centuries until the 1600s. India was left behind other rival seafaring empires, which made it easier for the British Empire to dominate India by the 1700s.

Christian empires were racing ahead in technological advancement, usually by borrowing technologies first invented elsewhere. From gunpowder and compasses to ink and paper, the characteristic features of European power were adapted from discoveries made by non-European cultures. Even the movable type method to print paper started in China. Gutenberg's improvements in the 1450s made the printed book simple to mass produce and distribute cheaply. Within decades of the invention of the printing press, many European city-states and entire nations passed strict laws about who could own a printing press, which books could be printed, how they could be sold, and where they could be transported. Think about how the construction and operation of a nuclear power plant is heavily regulated today, and you can better imagine how much control European monarchs demanded over printing presses. The mismanagement of a nuclear plant today is a frightening possibility, about as frightening as a renegade printing press seemed to a monarch or a pope centuries ago. The Roman Catholic Church updated its Index of prohibited books continuously from the 1500s until 1966, when it was finally abandoned. The authors who were listed included nearly all prominent philosophers who weren't Catholic, many scientists regardless of religion, and some great novelists as well.

Another innovation characteristic of modernizing Europe is the economic practice of money houses (we call them banks today) lending money at high interest rates calculated to reflect degrees of market risk. The sophisticated techniques of accounting and financial analysis had to be invented in the process. Loans at relatively low interest rates have been around as long as there has been coined money, of course. The financial relationships between the rich and the poor are far more influenced by substantial and long-term loans, thereby attracting religious scrutiny. Few major religions have omitted restrictions on lending from their concern for economic matters, especially when wealth is getting redistributed in the process. Major religions which have warned against charging high interest rates, then labeled as "usury," include ancient and medieval Hinduism, ancient Judaism, Islam, and medieval Roman Catholicism. The invention of mercantile capitalism didn't necessarily require high interest rates, but capitalism's explosive growth and evolution were materially aided where there were no laws against interest-bearing loans. Protestant countries had fewer religious concerns over loans at interest, which probably assisted the most rapid growth of capitalist innovations in northern European countries during the 1600s and 1700s. Our modern economic world was made possible by secularizing the economy away from religious oversight.

There is no single recipe for religious opposition to technology in general, and that is particularly the case with medical techniques. Two religions may happen to oppose the same technique, but on entirely different grounds. Many religions busily reinforce social customs that date far back before any memory or historical record, and many rigid customs are probably older than the religions enforcing them. Aside from Hinduism in India and Chinese folk religion, which are truly prehistoric, the major world religions around today are all less than 3,000 years old. With Hinduism and Chinese folk religion, the ancient doctrines and the ancient cultural ways are nearly impossible to distinguish from each other. With younger religions, their doctrines coincide with culture out of necessity, as they acquire plausibility in a particular geographical region by agreeing with that region's established cultural ways. Christianity and Islam, for example, harbor deep respect for peculiar values respecting patriarchy and authoritarianism that were common to the Middle East two thousand years ago.

A religion's moral expectations about gender, sexuality, reproduction, child rearing, marriage, family duties, duties to the elderly, rituals of death, and so on, all have their roots in cultural attitudes prevalent when its doctrines solidified. God's righteous commandments typically amounted to what the local populations already considered to be entirely right. The search for an established religion which abruptly disagrees with prevailing cultural standards will be fruitless, no matter where one searches all around the world throughout history. Lone religious prophets and reformers do erupt from time to time, managing to start new religions, but established religions are cultural entities first and foremost.

Religions have also invented arbitrary prohibitions against practices and technologies. There are no prohibitions against the use of condoms in the Bible, there is no commandment against abortion in the Bible, and there is nothing in the Bible forbidding genetic engineering on humans. The Bible doesn't mention any of them. The Bible couldn't speak to genetic engineering, of course, since its books are human-inspired and not God-inspired, and the science of genetics arrived long after the Bible. All the same, nothing in the Bible even vaguely warns against modifying our biological "nature."

Good Medicine, Bad Religion

There's no mention of any sort of birth control in the Bible, or in the Qur'an, or in the Hindu Vedas, or hardly any of the other scriptures of major world religions. Abortion isn't mentioned in the Bible either, so it wasn't specifically prohibited. Both the use of natural birth control techniques and pharmacological and surgical techniques of abortion were quite common across most civilizations 2,000 years ago. Since those primitive birth control techniques and those methods of abortion were crude, poorly applied, and prone to failure, they weren't encouraged very much, and most cultures back then tended to discourage them unless necessary. Ironically, many far more dangerous "medical" practices, based on sheer ignorance rather than knowledge, were tacitly permitted by religious scriptures. For example, some of the most stupidly dangerous techniques widely used by ancient physicians, such as bloodletting and herbal "remedies," aren't prohibited in the New Testament even though surely God could have realized by then that far more people were being killed than healed by them. Of course, this is the same God who approved of haphazard faith healing by an illiterate carpenter.

Religions blunder into discouraging the use of genuinely useful technologies, such as medical improvements, mainly due to their determination to reinforce many outdated traditional customs, by elevating them to religious rules. A good illustration is medical research into anatomy, to discover the body's organs and how they work. Although some ancient physicians did perform autopsies on deceased bodies, the practice faded out after Christianity dominated Europe. The attitude of Christian folk religion during medieval times coincided with the preference of the Church that the dead were not supposed to be violated so that a prompt burial, with all body parts in the correct place, could be honored. Between popular superstition and Church discouragement, no official Church decree against medical autopsy was needed; it was effectively illegal nearly everywhere in Europe until the late Renaissance. Other religions around the world have typically enforced prohibitions against violating the sanctity of the dead so that only the customary funeral practices would be respected. During the past two centuries, with the expansion of modern medicine, those traditional prohibitions have been eroding away. No major religion in the world today specifically prohibits autopsy, although smaller indigenous religions do tend to prioritize taboos about the dead.

A related medical technique is organ donation. Examples of traditional religions that discourage it include some Native Indian religions in North America and the Shinto religion in Japan. Christian Science discourages its followers from resorting to medical science in general. Jehovah's Witnesses rely on the Old Testament rule against eating blood—by taking a common sense idea out of its ancient context and inflating it to ridiculous proportions, the Jehovah Witnesses prohibit any sharing of bodily matter between people, including blood transfusions, based on flimsy scriptural interpretations. There is one Old Testament verse against eating blood in Leviticus 3:17, which also bans the eating of fat, yet Jehovah's Witnesses seem to be comfortable with its followers eating fat-marbled steaks. (Genesis does appear to command vegetarianism in verse 1:29.)

This recurring phenomenon of a new religion going out of its way to prohibit something "modern" can illustrate two of its features: it is rooted in some "authentic" past religion, and it can sharply separate its followers from others. A new religion needs both authenticity and distinctiveness in order to prevent its dissolution back into the environing culture over time. Followers have to be able to feel grounded in something original, and they have to feel special by comparison with everyone else around them to be uniquely blessed. Many religious prohibitions against modern technologies arise from these twin needs of new religions. The Amish are a paradigm of this. By repudiating most advanced technologies, they simultaneously drew a sharp line between themselves and the surrounding culture, and also made themselves heavily dependent upon each other in isolated communities. One additional factor was all that was needed to ensure their long-term survival: a migration to America's sparely populated farmland regions, to live where government prohibits religious persecution.

So far, we have discussed two bases behind religions enforcing their religious prohibitions against technologies: they are either reinforcing traditional customs as religious rules, or they are inventing new "sacred" rules just to maintain distinctiveness. There are two additional bases, to help account for religious prohibitions against technologies. The Christian discouragement of birth control serves to illustrate a third typical cause for religious prohibitions. Although there was no scriptural command or cultural tradition forbidding birth control in Europe, and the Church didn't pick condoms for prohibition to set up any arbitrary distinctiveness, this control over reproduction indirectly supported the rigidity of traditional customs. When a religion becomes concerned that some traditional customs need extra support and reinforcement, they invent prohibitions to indirectly serve that function. In the case of arbitrary religious controls over reproduction and pregnancy, which almost always fall on the shoulders of women rather than men, these religions are obviously reinforcing male dominance and authority over women in the family and the workplace. The religions manufacturing the "need" for these reinforcements invent some sort of theological justification, of course. The invented theological "reasons" needn't directly cite paternalism, but they might, such as claiming that God expects extra "protection" for women. Indirect support for paternalism can be phrased in terms of praising large families, or regarding fetuses as babies, or rejecting in vitro fertilization and test-tube babies. Whatever is so traditional about ways to subjugate women are often depicted as entirely natural—empowering technologies aren't "natural." Yet religions really don't care about what is actually natural. Science does.

Fetuses aren't babies, of course—few cultures confuse that obvious biological difference, and even fewer religious denominations, such as

Roman Catholicism since the 1860s, make dogmas out of conflating embryonic and fatal stages during pregnancy with babies after their birth. That's why such arbitrary and unsupported prohibitions, such as the prohibition of even early-term abortion, require completely manufactured "explanations" to bestow special status. Once a religion strays into the territory of blatantly constructing ad hoc religious reasons for arbitrary decrees, those theological excursions can only end up in fiction and propaganda. Theology must resort to indoctrination into propaganda just to get enough religious followers to go along with such a counterintuitive, untraditional, and arbitrary decree. The Catholic Church even had to resort to the Pope's newfound "infallibility" to establish its novel radical stance against all abortion. The theological propaganda that the Bible really regards all fetuses as babies only taught Christians how to misread their own scripture through distorting filters and then label that concoction as a "literal" reading.

Even worse, Christianity had to resort to retarding and distorting scientific knowledge which might contradict its decrees. This is an example of the fourth basis behind a religion's rules against a technology, where a religion feels compelled to deny science in order to prop up its arbitrary prohibitions. The Catholic Church had to resort to indoctrinating followers into fictions and lies, trying to forestall refutation by modern science. Embryology eventually verified many physiological and neurological differences between born babies and unborn embryos, more than sufficient to dispel confusions conflating them. Going all the way back in pregnancy to conception, there are even more physiological and genetic facts that deny any overlap between those embryonic beginnings after conception and the full development of a human person. Nevertheless, Christianity and other religions continue to distort, hide, and lie about scientific information concerning reproduction and pregnancy. This propaganda against science, which started from a need to prohibit a medical technology, erupted into a religious conflict against all of science. After one denomination of one religion, Roman Catholicism, came to the "rescue" of patriarchal domination with a bizarre decree about embryos, we are now in a situation more than a century later where religious people are all urged to regard science as their enemy. The "sacredness" of all human life and the "specialness" of our species as divinely created has aroused the Christian rejection of Darwinian evolution, a suspicion towards genetics, a fear of genetic experimentation and engineering, and a general attitude of hostility against the very medical technologies which are saving and extending lives.

These medical technologies can also make suffering people live longer than wise judgment would recommend. Most religions haven't felt it necessary to specifically discourage euthanasia as religiously forbidden. Many religions, including most denominations of Christianity, have come to some consensus about permitting the cessation of futile medical treatment after matters have gotten pretty hopeless. The distinction, however arbitrary, between "letting someone die naturally" and "prematurely killing someone" can still make a big emotional and moral difference to many people, both religious and secular. There are significant numbers of nonreligious people all around the world just as uncomfortable as religious people with the idea of a doctor deliberately killing a patient before that patient's illness would eventually be lethal.

Religions lacking a personal god usually maintain a prohibition against suicide, following long custom going back into the mists of prehistory. If euthanasia seems to look like a premature abandoning of hope and a violation of fate, there is religious disapproval against it. Most forms of Buddhism, for example, discourage euthanasia. Religions acknowledging a personal deity typically assign the responsibility for death's timing to a god. That divine responsibility seems to imply that we should not try to control death's timing. Oddly enough, modern religious people also expect doctors to control death's timing by delaying death by using medical technologies. Naturally, no one really notices this contradiction besides philosophers, along with doctors who have gotten annoyed by desperate families, who first say "God wants everything done for Grandma," and then later complain, "Doctors shouldn't play God" after things are going badly for Grandma. Any application of medical technology does involve controlling the timing of death, of course, one way or the other, considered from an objective standpoint. Religious people instead subjectively hope for a miracle, and when doctors manage to save a life with technology, declare how "God's work" has been done. But when that same medical technology is extending life past the point of any reasonable hope for a miracle, only then do religious people start thinking that the doctors are really at work here, doing something that God wouldn't do.

The theological contortions required for determining when God wants someone to die are not explainable in any rational manner. No religion has ever satisfactorily explained what to think about the one person surviving a plane crash, who promptly says that her god was "looking out for her that day." That lack of rationality infects any religious determination about when doctors are "killing" patients and when they aren't. So long as theological doctrine hasn't crushed a religious person's capacity to feel compassion towards a terminally ill and suffering patient, that compassion should speak loud and clear about the right thing that should be done to afford needed relief. Patients in such bad shape are usually receiving some medical treatments managing to keep them barely alive, so that removing those treatments, such as a respirator to assist breathing, is causing their death. Whoever is stopping the flow of oxygen to a patient is killing the patient, regardless of how that is accomplished. Saying that a patient has died "from natural causes" after having his respirator turned off implies that a person suffocated by a pillow died from natural causes, which in both cases is not enough oxygen in the lungs. Of course, the real difference isn't what is doing the killing, but why the killing is done.

A combination of cultural factors in the West has irrationally confused what is actually happening. In the West, Christian taboos against killing in general were extended to physicians long ago. During the twentieth century, political forces favoring patient rights established a legal right to refuse medical treatment. Doctors now were permitted to legally kill patients in a religious culture where doctors shouldn't kill patients. The American Medical Association followed the lead of some theologians who arbitrarily announced in the 1970s that only bad nature (or God) kills patients while good doctors merely "let patients go." Euthanasia is bad killing, so it never happens in hospitals, as the AMA self-righteously declared, and only "terminations of futile treatment" with good intentions do occur.

Secular people, regardless of their ethical judgments about how easily euthanasia should be available, don't need to suffer from distorted reality. Either doctors should be legally permitted to kill patients when there is excessive suffering, or some other medical professional should be assigned that technical task. Jack Kevorkian, the doctor convicted and jailed for administering a lethal injection in Michigan in 1998, suggested that licensed "thanatologists" should practice euthanasia and assisted suicide if doctors and nurses won't. Furthermore, on his view, assisted suicide should be legally available without anyone's having to feel responsible for causing a death.

If Dr. Kevorkian had lived in 1835.

Redesigning Humans

Maintaining and repairing our own bodies is nothing new. It isn't difficult to think of ways that our bodies could be better designed to handle
the stresses and strains of life. There's no divine designer to blame,
of course. Evolution's pressures on tree-dwelling monkeys produced
some primate species who dwelled mostly on the ground, and a few
species later explored the broad spaces between forests and savannahs
by walking on foot. Limbs and joints better adapted for climbing were
reshaped for walking and running, and necks and spines curved to bear
vertical weights. There's an ancestral reason why humans are always
complaining about sore feet and twisted ankles, bad knees and bad
backs, and neck and head aches. Surely there could have been a better
natural design, we might say to ourselves, as we nurse another nagging
injury.

We also wish for easier methods to repair our body parts after they
wear out. Humans have been inventing remedies and surgeries at a
faster and faster pace since the dawn of civilization, but the limitations
of our natural design somehow stay ahead of our efforts. Organs degenerate too quickly, vital systems clog too much, and even our cells are
programmed for certain death. Death is as natural as life, and efforts to
extend life are no less natural than either. Claiming that extending life
is unnatural is ridiculous, since life itself is a long story of experiments
with extending life beyond the fast-expiring chemical reactions of simple organic compounds. Our bodies are the result. All of our cells are
complex factories with many parts, and we are multicellular organisms
with many organic components, precisely because evolution has randomly had many successful experiments for extending life. Our species
no longer suffers from a lifespan of less than 30 years, because our
natural talents for inventing technologies have saved lives that would
have been lost and sustained health that could have been squandered.
Extending life, in itself, is no more unnatural than multicellularity or
technology.

We enjoy the technologies familiar to us, but new technologies can
seem abrupt and surprising, an unnatural deviation from the familiar and comfortable. Someone quite happy to wear glasses and drive

a car may find a prosthetic hand disturbing. It's a common enough bias. What was accepted during one's upbringing feels entirely normal, and hence natural, but as one's years lengthen, anything too deviantly novel is taken to be abnormal, unneeded, and unnatural. These intuitive judgments can be dressed up into arguments with premises, but they rarely amount to anything more than rationalizations expressing an instinctive repulsion against the unfamiliar. The most philosophical of antitechnological arguments offer nothing more complicated than statements of fear, resentment, or resistance.

What do we typically hear from naysayers? Complaints revolve around the same basic anxieties. We can easily imagine what is really getting said through their words. "They shouldn't be doing that!" Perhaps the plain meaning goes something like this: If too many people start doing things that new way, then not enough people will still do things my way, and that feels threatening. Or perhaps the actually worry is this: If so many people are more like that, many people will still remain like me, and that feels unfair. The real issue could also be: If so many people become that way, next they will expect me to join them, and that feels coercive. These are honest and revealing sentiments. But sincere disapprovals are nothing more than expressions of negative emotions, rather than thoughtful deliberations. Humanity cannot afford such irresponsible and selfish negativity, and has never listened for long, anyway. If humanity had been effectively restrained by all that apprehension, we would still be huddled around small fires under dark skies with snarling predators closing in.

There is nothing more human than the inventive quest for improvement. Improvement by experiment is never a matter of safety or equality. Some must bravely develop new capacities so that others may subsequently imitate the boldly successful. Redesigning human beings will cause them to do things differently and live differently, in many ways beyond prediction—and indeed, that is always the whole point. Complaining that the consequences of a human redesign cannot be entirely predicted is a pointless cry of alarm, since we always make bold ventures without knowing everything to come. Indeed, that is the very essence of exploration. The unexpected can often be, on the whole, more valuable than the expected. Predictably bad consequences, even if they have a small chance of occurring, should be wisely considered and weighed against good results, no doubt. Precautionary rules are never out of place, so long as precaution isn't just a pretext for emotional resistance.

Using technology to improve our biological selves is just the beginning. Computerized technologies busily repairing our organic systems will inevitably lead towards the blending of humanity and machinery. Shall humanity merge with computing to become partly cybernetic and robotic? Shall we decode the instructions of the genetic code of DNA and RNA to the point where adjusting that genetic recipe could reliably improve our design? We have already embarked on hesitant steps towards these goals, with some preliminary yet encouraging results. Are we leaving behind Humanity version 1.0 in order to invent an upgrade, as sociologist Steven Fuller suggests? His recent book *Humanity 2.0* is a fine example of a growing genre of books cautiously optimistic about the beneficial opportunities beckoning us into the future. As a philosopher of technology, he is well aware how humanity has made several vast leaps of technological progress to get where we are today. A wider perspective could assign the label of Humanity 5.0 instead, if skills with stone tools (1.0), the use of fire (2.0), the Neolithic arts of agriculture and construction (3.0), and the mastery of electromagnetism (4.0) are given proper respect for forever altering how humanity lives. The next major step, aligning the digital flux of the electron with the metabolic machinery of the cell, will stand among those previous evolutionary leaps.

With modernism comes new monsters. Not the fabled creatures of legend like the werewolf or the vampire, but monsters we could make ourselves. The animated Golem, the Frankenstein creature, the chimeric Hybrid, the sentient Robot, the Android, the Cyborg—each age generates its characteristic monstrous figures, vaguely made in our image, yet so different from us. Our fear of our monsters is just the outer layer of worry. The deeper anxiety is that we might permit ourselves to become those monsters, too. The fear of the monstrous isn't the fear of the completely alien, but rather the fear of the slightly inhuman, sharing our worst traits while displaying few of our virtues. Our first concern about these near-human monsters is that they wouldn't behave as morally as we do, causing even more problems in the world than we humans already cause. On deeper reflection, our more fundamental concern is that they couldn't even recognize morality at all, or that they would have their own sense of morality quite different from ours. They couldn't regard us as part of the same ethical community—as eligible to be given the moral respect and concern as they give each other—so that we would end up looking inferior and unimportant in their eyes. Just as we now treat the other animals around us, as protected pets, unwanted pests, or likely lunch, would these monsters similarly feel free to do whatever they want to us?

These serious concerns can add up. Anxiety over whether some humans might become monsters could be joined by the worry that those monsters consider humanity to be beneath them. Humanity could divide into two species without any common ethics uniting it anymore. Since the whole point of ethics is attempting to hold all humanity together as a moral community with some common rules or ideals, creating such monsters amounts to the destruction of ethics. Nothing could be more unethical than that. This judgment upon near-human creatures is philosophically sound, and ethically wise. We already live in a world where the simple bonds of moral community are too fragile. Slavery hasn't been eradicated yet, tyranny still flourishes in regions of perpetual poverty, and the richest countries do little for the poorest. Looking at so much shameful inequality, one could conclude that the truly unethical thing is to keep encouraging those who have so much to accumulate even more, and leave everyone else even further behind. There is a building resentment against amazing technology permitting a few to live so much longer and better while so many are mired in destitution and decay. There is a growing suspicion against advanced technology when its promoters speak of improving humanity while everyone knows that only perhaps 20% of humanity will actually get access.

A dystopian future is easy to envision, because it is easy to get there: the superhumans don't really notice how far ahead they are, because their improving lives have quietly enjoyed steady acceleration away from the rest of humanity. When you can take perpetual progress for granted, you can only be astonished when you finally turn around to look back at how far you have traveled. Most of humanity is living in that dystopian nightmare right now, mired in terrible poverty, while 10% of humanity became incredibly healthy and wealthy over just the past 10 generations. That fortunate 10% have become monstrous, manipulating economies and governments around the world for their own selfish benefit, and waging wars at their pleasure on other peoples' homelands. The worldly ethics of this top 10% have proven to be quite flexible, making sure that dutiful emotions of caring and compassion get piously pronounced, while requiring little in the way of civic action to maintain social justice in reality.

Religions, for their part, have also been ethically "flexible" in response to the vast disparity between the "haves" and the "have nots." The religious denominations preferred in the wealthiest countries have very little to say against the accumulation of wealth, lending comfort to those who want to suppose that god most appreciates the upper class and its controlling powers. By contrast, many religious denominations preferred in poor countries either align god with preserving the traditional ways of peoples standing united, or they depict god as personally concerned for the struggles of individuals trying to better themselves amidst the chaos around them. The same religion can sometimes offer all three types of denominations simultaneously. Christianity, for example, has varieties of Protestantism appealing to the comfortably wealthy (the "mainline" established churches), and other varieties appealing to the hopeful poor (the "evangelical" churches), and yet more denominations preserving communal unity around conservative values (for example, Roman Catholicism). Buddhism has exemplified these three primary types of denominations as well across different countries over many centuries. Islam is a younger religion, yet it has displayed variety in its past as well, and it is showing signs that it will be more than just a conservative social force into the future. Yet Islam happens to be playing the role of a deeply conservative and antimodernism force in many countries today.

If god is ultimately responsible for nature, and nature's ways signal what is truly good, then religions can easily appeal to what is "natural" in order to reinforce values. But what really is "natural" for human beings? There is no consensus anymore, if there ever was. Any belief system can simply announce what it thinks is most natural, and pass ethical judgment accordingly. (Worldly ethical systems can do this too, by simply omitting god, and letting their conception of nature play the role of setting moral standards. Better philosophical ethics wouldn't ground morality so simplistically, however.) Nature never settles matters, though. Different religious denominations end up disagreeing about what is really so "natural." If it is assumed to be quite natural for individuals to improve themselves and accumulate wealth, then god must approve, and hence anyone skeptical about wealth is questioning god's wisdom. If it is instead most natural for people to endlessly modify environments around them, then god must approve, and anyone calling for less exploitation of nature is going against god's will. Today, we can even ask questions about people acquiring the ability to modify themselves, with surgical, medicinal, or genetic enhancements. Is it "natural" for people to take advantage of technology to repair and strengthen their bodies? Few religions have rejected medicine entirely, proving how common sense usually prevails over religion. Advanced biotechnologies continue to prevent tragic death and extend our healthy powers, and common sense won't reject those beneficial improvements either.

Common sense is usually the best antidote to any religious or worldly ethical theorizing that presumes to figure out when enough technology is enough. Religion is particularly useless. Religious ethics is either hopelessly tied to dogmatic scripture from some past age, or arbitrarily selective about what should be taken to be "natural." For example, a religion hostile towards genetic modification to human DNA, on the grounds that human DNA should remain "pure" and unchanged from its original "design," is oblivious to scientific reality. There is no such thing as "pure" or "original" DNA, however, since every human being is a unique organism full of random mutations and accumulated genetic accidents. At what point could any DNA be called "original"?

For example, much of the DNA inside any mammal consists of left-over genetic material from viral invasions and bacterial interventions from a deep ancestral past, as animals evolved over hundreds of millions of years. Further genetic combinations and mutations eventually produced our species, so we are all beneficiaries of ceaseless genetic change.

If we take even more control over the human redesigning of our evolutionary inheritance, we couldn't be doing anything "unnatural," since nature has been continually redesigning us without rest. Enhancement is the story of life itself. The only question is whether we will do what nature does with greater intelligence and enhanced wisdom.

Eugenics

Worldly ethics can fall into perplexities and paradoxes just as quickly as any religion, if it is naively imagined that "the natural" can serve as reliable guidance about ethical choices. Does science's knowledge of nature directly provide moral knowledge too?

The last significant eruption of naturalistic ethics, actually making a major civic and political impact, was Social Darwinism, during the late nineteenth and early twentieth centuries in many countries such as the United States and Germany. If Darwinian evolution seems to say that nature expects only the fit and strong to survive—this is already a distortion of biological evolution—then the fit and strong should be encouraged to survive even better! But who are the fittest now among human societies? Asking this question invites further distortions to biology. The criteria we now use for human "success" look little like the qualities that allowed Homo sapiens to survive in Africa after it had evolved. Ignoring scientific biology and what Darwin himself wrote in his books, wealthy people feeling successful and powerful took the initiative answering that question, and they said, "We must be the best!" Social Darwinism promptly declared that the most fit people are the most wealthy, so therefore the rich deserve their wealth, while the unfit poor should die off. If the unfit aren't dying off fast enough, then the civic programs of Social Darwinism could speed things up, according to its proponents. Those programs relied on sheer pseudoscience, not anything that even the biology of those times could authorize.

I did NOT say that!

The science of eugenics was also established by the mid-nineteenth century to figure out how to apply biological and medical technologies to prevent the birth of "unfit" children and promote the birth of "more fit" children. Much evil has been done in the name of eugenics, as well. Only careful ethical thinking can discriminate what is morally acceptable, and what must remain unacceptable, about eugenics. A few modest proposals that could be classified as "eugenics" may survive that rigorous ethical scrutiny.

Social Darwinism is an entirely ugly ideology, by contrast, underserving of the label of "ethics." Not without reason is Social Darwinism widely recalled as a dangerous worldview. Many religious people suppose that Social Darwinism is precisely what must follow from naturalism. Fortunately, that is hardly the case. Although everything about humans is all natural, it does not follow that nature fully determines the good life for humanity. We are already naturally unusual, after all—our evolutionary and genetic pathway isn't the same as any other earthly organisms. We are earthlings, but we are also humans, with our own journeys as a special species. The record of who we have been and what we have achieved is archived in our cultures, as well as our genes. Social Darwinism ignores the history of human cultures, forgetting how cultures are repositories for tried and tested wisdom about better ways to live and prosper. Without trying to understand human culture, Social Darwinism tried to foresee the course of human evolution, picking future winners and losers. By its calculations, individual competition without any rules would naturally sort out the winners. No wonder only immorality could be the result. Morality isn't in "natural" interactions among people living on their own, but in cooperative social relationships among people working together.

Common sense and sincere compassion, as cultures around the world agree, supply better guidelines for life than mere competition. Common sense can reliably tell when morally wrong things should be prevented, such as tragic deaths and avoidable harms. Heartfelt compassion, combined with sound wisdom, points out injustices that leave some too far behind while others race too far ahead. Common sense morals and compassionate justice can do plenty of work sorting out which technologies should be used now, which shouldn't be used now, and which need to be developed in the future. Does a new biotechnology prevent suffering and improve health? Does a new biotechnology provide the whole human community with more security, fairness, and cohesiveness? Those who expect futuristic technologies to make humanity far better must be the most responsible for ensuring that our humanity isn't lost in the process.

Crucial questions about ethics, rather than speculative questions about utopias, need to be asked first. "Eugenics" has acquired a negative connotation, but preventing the birth of irretrievably damaged or diseased fetuses is positively moral. Biotechnology arouses fears of monsters, but if we limit human modifications to those which enhance communal solidarity and promote rising living standards for everyone, modified

humans will be exemplary leaders instead of fearsome tyrants. There are naturally wise and humane tests for exposing the experiments too unethical to risk, even if we don't agree about all moral matters.

As for religious objections to biotechnological improvements to humanity, they can all be safely ignored. They either amount to ignorant or unscientific notions about what counts as "natural," or they point to divine commands about what conveniently preserves traditional privileges and prejudices. Have religious objections to reproductive technologies such as birth control or in vitro fertilization really been about preserving natural dignities or rights, or actually about protecting traditional male privileges and controls over women? Are religious worries over embryonic genetics and stem cells really about elevating the intrinsic ethical status of every human being, or really about perpetuating cultural battles over access to medical procedures such as abortion or genetic testing? Religions losing political control over everyone's lives eventually resort to last-stand ethical battles where they demand the last moral word. The modern world can determine how best to improve peoples' lives without interference from humanity's remaining tyrants.

Where religious concerns do manage to appeal to genuinely ethical worries, we can all consider those objections with ethical standards, rather than unnecessary dogmas, foremost in our minds. In general, when religious voices stand opposed to new biotechnologies in the name of "human dignity," we should all wonder why religions have fought so hard for so long to keep humanity ignorant, fragmented, and fragile. If humanity bereft of science and medicine is religion's vision of a dignified human life, then we can dismiss such pious protests with a clean conscience.

A Planet Worth Saving

The ethics of human enhancement should be guided by commonsensical and compassionate wisdom, so that a genuine "enhancement ethics" can lead the way. While we have been busily modifying our local environments and even our own bodies, we have also been causing immense changes to the entire planetary biosystem. We now need a genuine "ecological ethics" as well.

Religions have been hopelessly left behind, for the most part, especially their dogmas about some god who is ultimately responsible for the world. If a supernatural deity is intelligently controlling the world, then whatever happens here is proceeding according to divine wish or plan. Unfortunately, a god focused on saving some for an eternal life after this world won't be prioritizing what happens to this world. If no god is caring for this world in the long run, then as far as this worldly life is concerned, we might as well suppose that we are on our own.

Yet, is humanity so alone? There is plenty of life all around us, after all. If we take a fresh perspective on our kindred relationship to all earthly life, then we should feel a renewed sense of responsibility towards each other and all living things on this planet.

No god, supernatural or natural, will save us from the fate we deserve if we destroy the only home we will ever know. Our fate may be closer than we suspect. Our moral standing with the planet couldn't be lower, as we parasitically drain the vitality from it. Polluting every habitat and ecosystem as fast as we can is not a sustainable plan. Some people fancifully hope that a loving Earth will repair itself and us. But not even a planet can keep pace with a rapacious species like humanity. We aren't that different from any other species, in a way. The number

of species that have voluntarily used fewer resources, to date, is precisely zero. The difference with humanity is the way we exploit technology to exhaust all resources at an exponential rate. Technology has been our destiny, and it will determine our fate.

It is unlikely that humanity will do anything else besides applying more technology, from new energy sources like fusion to novel ways to wage war, to deal with its problems. Hope for humanity's survival may rest on further technological advancement. Unless physics is misguided about relativity and space-time, travel to other stars' habitable planets within an astronaut's single lifetime will not happen soon. For small colonies of star voyagers willing to take generations to journey to a new home, there probably are a few nearby planets roughly similar to our own. With advanced technologies, humans could slightly modify planetary temperatures and air pressures and adjust the atmospheric conditions to make them habitable. We are already in the process of discovering some Earth-like planets fairly close to home. One example is Gliese 581g, only 20 light years away, which is a short distance considering the vast size of the Milky Way galaxy by comparison.

What is this hope for new planets to populate, but an expression of that anciently human drive to explore and exploit fresh territory? There are other, far closer options. Any technology sufficient for star traveling and planet terraforming would be sufficient to clean up this planet we already have. We might use novel technologies to evolve into indestructible cyborgs, merging humanity and machinery. We may create artificially intelligent machines that are capable of replicating themselves, machines that far exceed our current human intelligence, machines that could continue to live and thrive by withstanding the onslaughts of our violent universe.

Our species is already highly evolved compared to others on the planet, so much so that we have become agents of change just like the mother nature. It is becoming apparent that we are the very first species ever to evolve on this planet that will decide its own evolution. We are about to become a much bigger force of nature. Humans are presently altering the Earth's land, water, and air faster than any competing geologic process. Although we can be proud of ourselves for our ability to become a powerful force of nature, we also need to be embarrassed when we use this ability in transforming our watersheds, leveling beautiful mountains, eliminating dense forests, changing the balance of gases in our atmosphere, and extinguishing thousands of other species—destroying all the things that have sustained and nourished life throughout the history of our planet.

Science has revealed the incredible journey that our home planet has taken from its fiery birth to today. We live on a warm, watery, and living globe. As soon as life could get started it spread everywhere, survived many challenges, and transformed the Earth's surface in ways to permit its continued evolution. Feeling deep respect and reverence for our home, so far our only home, is entirely natural. Now we must act dutifully, and act quickly, from that sense of profound reverence.

Humans, as well as all other species that ever existed, have survived entirely at the mercy of our planet and other natural forces prevailing in our solar system. We know that more than 99% of all species that ever lived have gone extinct. It is difficult to believe that we would be an exception—species eventually fall victim to something, an asteroid, a super volcano, a gamma ray burst, a disease, a famine, etc. Homo sapiens, with its vast intelligence, has aroused additional risks to itself and the entire planet, such as nuclear war, poisoned oceans, depleted drinkable water, and global warming.

It is hard to understand how we can destroy everything that has sustained our lives and yet expect to continue to live as a species here. Humanity had better understand the nature and purpose of our own earth-shaping powers if we are to have any hope of controlling them. Life is stubborn and persistent, made for survival. If our damage to the planet gets beyond our control, life will still go on, long after humanity is gone. The earned pride we can take in coming so far to occupy where we stand now will melt away from the shame of having thrown it all away. If humanity's story is essentially about replacing ignorance with knowledge, let us use that knowledge responsibly now that we possess it.

Further Reading

Archer, David. *The Long Thaw: How Humans Are Changing the Next 100,000 Years of Earth's Climate.* Princeton, N.J.: Princeton University Press, 2008.

Bannister, Robert. *Social Darwinism: Science and Myth in Anglo-American Social Thought.* Philadelphia: Temple University Press, 2010.

Bryant, Clifton D., ed. *Handbook of Death & Dying.* Thousand Oaks, Cal.: Sage, 2003.

Dowbiggin, Ian. *A Concise History of Euthanasia: Life, Death, God, and Medicine.* Lanham, Md.: Rowman & Littlefield, 2007.

Egerton, Frank N. *Roots of Ecology: Antiquity to Haeckel.* Berkeley: University of California Press, 2012.

Fuller, Steve. *Humanity 2.0: What It Means to Be Human Past, Present and Future.* London: Palgrave Macmillan, 2011.

Goldberg, Michelle. *The Means of Reproduction: Sex, Power, and the Future of the World.* New York: Penguin, 2009.

Greene, Rebecca. *History of Medicine.* London and New York: Routledge, 2013.

Headrick, Daniel R. *Technology: A World History.* Oxford: Oxford University Press, 2009.

Johansen, Bruce. *The Global Warming Combat Manual: Solutions for a Sustainable World.* Westport, Conn.: Greenwood Publishing Group, 2008.

Kevles, Daniel J. *In the Name of Eugenics: Genetics and the Uses of Human Heredity.* New York: Random House, 2013.

Lilley, Stephen. *Transhumanism and Society: The Social Debate over Human Enhancement.* Berlin and New York: Springer, 2012.

Matossian, Mary. *Shaping World History: Breakthroughs in Ecology, Technology, Science, and Politics*. Armonk, N.Y.: M.E. Sharpe, 1997.

McClellan, James E., and Harold Dorn. *Science and Technology in World History: An Introduction*, 2nd edn. Baltimore, Md.: Johns Hopkins University Press, 2006.

Moazam, Farhat. *Bioethics and Organ Transplantation in a Muslim Society: A Study in Culture, Ethnography, and Religion*. Bloomington: Indiana University Press, 2006.

Nicol, Neal, and Harry Wylie. *Between the Dying and the Dead: Dr. Jack Kevorkian's Life and the Battle to Legalize Euthanasia*. Madison: University of Wisconsin Press, 2006.

Pickstone, John V. *Ways of Knowing: A New History of Science, Technology and Medicine*. Chicago: University of Chicago Press, 2001.

Rosen, Christine. *Preaching Eugenics: Religious Leaders and the American Eugenics Movement*. Oxford: Oxford University Press, 2004.

Sands, Kathleen M. *God Forbid: Religion and Sex in American Public Life*. Oxford: Oxford University Press, 2000.

Silver, Lee. *Challenging Nature: The Clash between Biotechnology and Spirituality*. New York: HarperCollins, 2006.

Tobin, Kathleen A. *The American Religious Debate over Birth Control, 1907–1937*. Jefferson, N.C.: McFarland, 2001.

Vasey, Daniel E. *An Ecological History of Agriculture, 10,000 BC–AD 10,000*. Lafayette, Ind.: Purdue University Press, 2002

Whittaker, Andrea, ed. *Abortion in Asia: Local Dilemmas, Global Politics*. Oxford and New York: Berghahn Books, 2012.

Wimberley, Edward T. *Nested Ecology: The Place of Humans in the Ecological Hierarchy*. Baltimore, Md.: Johns Hopkins University Press, 2009.

INDEX

ABOUT THE AUTHORS

Paul Singh is professor of Obstetrics and Gynecology at the College of Medicine, University of Science, Arts and Technology at Montserrat, British West Indies. He is the founder president of Singh Global Initiatives, a philanthropic organization that promotes health and science education worldwide.

John R. Shook is a scholar and professor living in the Washington, DC area. He received his PhD in philosophy from the University of Buffalo and has authored or edited more than a dozen books, including *The God Debates: A 21st Century Guide for Atheists and Believers (and Everyone in Between)*.

OTHER TITLES FROM PITCHSTONE

From Apostle to Apostate:
The Story of the Clergy Project
Catherine Dunphy

God Bless America:
Strange and Unusual Religious Beliefs and Practices in the United States
by Karen Stollznow

How to Defeat Religion in 10 Easy Steps
by Ryan T. Cragun

Life Driven Purpose:
How an Atheist Finds Meaning
by Dan Barker

A Manual for Creating Atheists
by Peter Boghossian

The Necessity of Secularism:
Why God Can't Tell Us What to Do
Ronald A. Lindsay

PsychoBible:
Behavior, Religion & the Holy Book
by Armando Favazza, MD

What You Don't Know about Religion (but Should)
by Ryan T. Cragun

Why Are You Atheists So Angry?
99 Things That Piss Off the Godless
by Greta Christina

Why We Believe in God(s):
A Concise Guide to the Science of Faith
by J. Anderson Thomson, Jr., MD, with Clare Aukofer